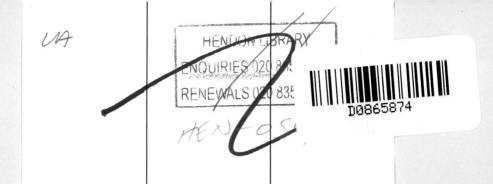
Please return/renew this item by the
last date shown to avoid a charge.
Books may also be renewed by phone
and Internet. May not be renewed if
required by another reader.

www.libraries.barnet.gov.uk

BARNET
LONDON BOROUGH

200

Home-made Treatments
for Natural Beauty

200

Home-made Treatments
for Natural Beauty

Shannon Buck

APPLE

A QUARTO BOOK
Copyright © 2014 Quarto plc

Published in 2014 by
Apple Press
74–77 White Lion Street
Islington, London
N1 9PF

www.apple-press.com

ISBN: 978-1-84543-545-5

Conceived, designed and produced by
Quarto Publishing plc
The Old Brewery
6 Blundell Street
London
N7 9BH

QUAR.HBT

Project editor: Lily de Gatacre
Art editor and designer: Julie Francis
Copy editor: Caroline West
Photographer: Simon Pask
Illustrators: Tracy Turnball and Juliet Percival
Design assistant: Martina Calvio
Proofreader: Claudia Martin
Indexer: Helen Snaith

Art director: Caroline Guest
Creative director: Moira Clinch
Publisher: Paul Carslake

Colour separation in Singapore by Pica Digital Pte Limited
Printed in China by 1010 Printing Limited

10 9 8 7 6 5 4 3 2 1

contents

Foreword

I became interested in natural beauty in my mid-twenties, developing my own formulas in my kitchen and finding them significantly more effective than the high-priced products that I would waste my money on during school and university. I still can't believe that I used to shell out good money on all the hyped-up and ineffective wrinkle-erasing, lip-plumping, pore-reducing, chemical-containing conventional beauty potions with extravagant labels and outrageous price tags.

When my son was a baby, he had extremely sensitive skin and was prone to severe eczema and rashes. After some research into essential oils and herbal medicine, and using only the restorative powers of natural ingredients, I developed formulas that worked wonders for my baby's precious skin.

I am a professionally trained herbalist and studied Aromatherapy & Essential Oils at Bastyr University. I also teach an introductory course on Aromatherapy and Essential Oils, and in 2011, I launched a beauty blog called FreshPickedBeauty.com where I create step-by-step tutorials for creating your own natural skincare recipes. My blog has turned out to be quite popular and is visited by thousands of fantastic people from around the world every day.

I know you, too, will take pleasure in discovering the fun of formulating your very own natural beauty recipes. Green blessings!

Shannon Buck

About this Book

In the pages of this book you'll find more than 200 expert tips, hints, recipes and problem-solving solutions that make creating customised natural and organic beauty products in your own kitchen easy and enjoyable. The book is divided into six chapters covering everything from hair care to pedicures so you can pamper yourself from head to toe.

1 Selecting Ingredients pages 8–41

In this chapter is all you need to know about the natural ingredients that will be at the heart of your beauty products. From sugars and salts to floral waters and decadent essential oils, you'll find information on sourcing and storing your organic ingredients as well as wonderful ideas for how to use them.

2 Tools and Supplies pages 42–51

Almost everything you need to get started is right in your kitchen already and this chapter will guide you through selecting the right tools and equipment, as well as giving you crucial tips on safety and sanitation. Learn how to select eco-friendly packaging for your finished products and get some ideas for how to package your creations to make gorgeous gifts.

3 Naturally Lustrous Hair pages 52–71

Whether your hair is long or short, dry or oily, straight or curly, you'll find great tips and luxurious recipes here to cleanse and condition your hair like never before.

4 Fabulous Facial Care pages 72–101

Every aspect of facial care is covered in this chapter, from toners and moisturisers, scrubs and masks, to specialist eye and lip care. These tips and recipes will leave your complexion blemish-free, hydrated and radiant.

5 Conditioning Body Care pages 102–125

Pamper your whole body with the treatments in this chapter: soak your cares away in an indulgent bath tea, create your own natural deodorant and treat yourself to the perfect pedicure.

6 Perfumes and Aromatherapy pages 126–137

Enter the intoxicating world of aromatherapy and discover how to blend essential oils to treat your body and your home, with perfumes, massage oils and air diffusers.

The Best Recipes

Here you'll find the author's top recommendations for wonderful products for you to make at home, complete with ingredients list, skin-type recommendations and usage advice.

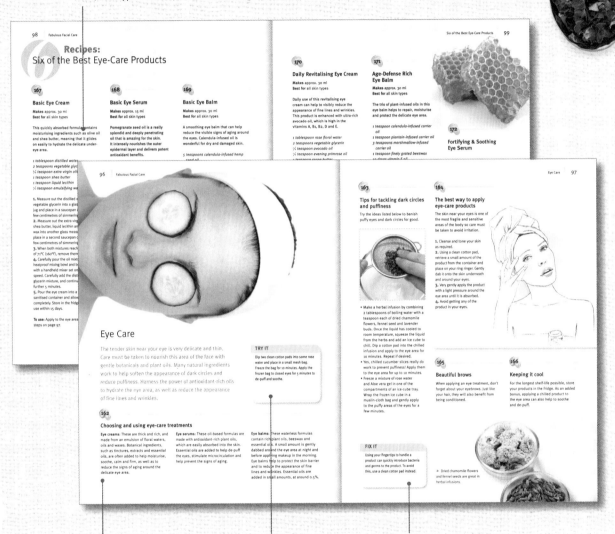

Tips

From advice on selecting and storing ingredients, to making and customising treatments and the best way to apply products to your skin, the book is full of expert tips and advice to give you guidance and inspiration.

Try Its

Try something new, discover a quick fix to a beauty bother or learn a new time- or money-saving trick with these useful panels, which appear in every chapter.

Fix Its

These pop up throughout the book and offer great problem-solving hints to help you avoid common pitfalls or repair products that have gone wrong.

1 Selecting Ingredients

Shop-bought beauty products are often laden with synthetic ingredients, fake fragrances, unnatural fillers and irritating chemicals. The best thing about creating your own beauty products is that you can know exactly what you're putting on your skin and feel confident that it is nourishing and pure. At the very heart of your handmade beauty products will be the natural, organic and wild-harvested ingredients that you select. This chapter will guide you through choosing and using the absolute best ingredients that Mother Nature has to offer.

Buying Natural Ingredients

These days it is pretty easy to find natural, organic and wild-harvested ingredients to use in hand-crafted skincare formulas. Here are just a few of the reasons why you should use these natural ingredients.

1

Grow them! Wild-harvest them! Buy them!

If you choose to cultivate your own organic plants and herbs, wild-harvest them considerately from sustainable sources, or buy them from a highly regarded natural products company. Then you can feel good about your decision to use the finest available ingredients. You should procure many of the ingredients, such as essential oils, carrier oils, butters, waxes and floral waters, from dependable companies. Here are some significant features to look out for when deciding which company to purchase from:

- Does the company offer mostly, if not all, organic and/or wild-harvested products and ingredients?
- Is the company a certified organic processor or fully accredited with the Department for Environment, Food and Rural Affairs (DEFRA)?
- Does the company exercise Fair Trade practices when dealing with the growers and harvesters of their ingredients and products?
- Is the company mindful of the environment and does it use recycled materials, post-consumer waste materials, soya-bean-based inks and non-toxic chemicals in printed materials such as flyers, brochures, catalogues, and receipts?
- Does the company use sustainable packaging for all of its products? Glass packaging is best, although the use of PETE (polyethylene terephthalate) is a practical recyclable plastic option.
- Does the company offer fresh products at fair prices?

Why choose natural?

- You can avoid putting ingredients that have suspected health risks on your body, which can then be absorbed into your system.
- When ingredients are grown organically and naturally, they are healthier, more nourishing and ultimately more beneficial to your body.
- You are supporting organic agriculture, your local farmer and Mother Nature.
- You are supporting a healthy ecosystem by keeping toxic chemicals out of the soil, water and air.

Some helpful definitions

Please note that these definitions may differ throughout the world.

100% Organic: The product contains only organic ingredient(s), and no pesticides or fertilisers were used during the growing process.

Certified Organic Retailer: The retailer of the ingredients follows strict rules regarding how it handles, stores and sells its products.

Made with Organic Ingredients: The product is made with at least 70% organic ingredients and is certified according to national organic standards.

Natural Ingredient: The ingredient came from or was made from a renewable natural resource without any petroleum compounds or synthetic silicone.

Organic: The product contains at least 95% organic ingredients and is certified according to national organic standards.

Organic Crops: The crop was grown without the use of irradiation, harmful pesticides, genetically modified organisms (GMO), sewage sludge and synthetic fertilisers.

Wild-harvested: The natural ingredients were harvested from their wild habitat. Consideration is given to harvesting only what is necessary so that plant and animal species living in the same habitat are not harmed.

Safety and Patch Testing

Most of the ingredients and methods described in this book for creating your own kitchen-crafted skincare products have been used successfully for countless numbers of years. Most are known to be safe and beneficial to the skin and body when used externally.

Your skin is the largest organ in your body and will absorb some of what is applied to it. Just as you may choose to eat fresh and organic food, you will also feel happy choosing only organic and fresh ingredients when making your own beauty products. By doing so, you will be able to avoid introducing damaging and synthetic substances into the body via your skin. When you are sourcing natural ingredients to use, it is strongly advised that you look only for those certified as organic or ethically wild-harvested.

Even though you'll be using natural and/or organic ingredients in the recipes in this book, bear in mind that you will need to avoid certain ingredients due to any of the following conditions:
• If you are allergic or sensitive to a particular ingredient.
• If you are pregnant, wish to become pregnant or are breast-feeding.
• If you have other health conditions that require consultation with a healthcare professional.
• If you are on any prescribed and/or over-the-counter medication.

Note: Consult a healthcare professional before using any ingredients recommended in this book if necessary.

How to perform a patch test

Before you apply or use any ingredient that you have never used before, you may wish to perform what is commonly called a 'patch test' to determine how your skin will react. This is particularly important when using essential oils. Here is how to perform a patch test:

1. If you are testing an essential oil, dilute two drops of the oil in one teaspoon of jojoba oil. Other ingredients may be applied undiluted. Using a cotton bud, apply a small amount of the ingredient to the inside crook of your elbow and leave on for 24 hours without washing off.

2. Monitor the area for an allergic reaction or sensitivity (such as pain, bumps, redness, rash, itchiness or other changes). If you develop any of these symptoms, then see your healthcare professional before continuing to use the ingredient.

3. If you do not experience an allergic reaction, it is probably safe to continue using the ingredient. However, skin can change over time for many reasons. Therefore, if you believe that you are having an allergic reaction to an ingredient in the future, stop using the ingredient and seek the advice of your healthcare professional.

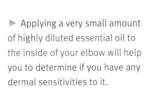

▶ Applying a very small amount of highly diluted essential oil to the inside of your elbow will help you to determine if you have any dermal sensitivities to it.

Natural Butters

Natural butters are expeller-pressed from the seeds and kernels of trees, and are solid at room temperature. Butters are employed extensively in beauty recipes to impart a creamy, smooth and dense consistency to lotions, creams, lip balms and even soaps. Butters are truly splendid and fantastically pampering, and frequently used alone to condition and care for the skin.

Naturally sourced butters vary in hue from white, through off-white, creamy coloured and pale yellow, and even to tan or a greyish colour. They can be purchased refined or unrefined, with consistencies ranging from soft to semi-hard to very hard.

'Unrefined' means that the butter did not pass through a filtering system and/or was not treated with any chemical or solvent to alter its colour, texture, aroma, vitamin content or natural properties. It is common to find 'refined', 'ultra-refined' and 'deodorised' butters for sale. If you truly desire the butter in its natural form, choose a raw or unrefined version.

Natural butters are recommended if you are making lotions, creams, body butters, lip balms, lotion bars and conditioners.

3

Choosing the best butter for different recipes

Cocoa butter (1): A creamy to pale-yellow-coloured semi-hard butter that is expeller-pressed from the seeds of the cacao tree (*Theobroma cacao*). Cocoa butter imparts a delectable chocolate aroma to your skincare product. It is a magnificent ingredient to use if you want to reduce dryness and improve the elasticity of your skin. Cocoa butter is frequently used in formulas that support a reduction in the appearance of stretch marks. You can purchase it in convenient small wafers to assist with easier melting.

Kokum butter (2): A white-coloured hard butter that is expeller-pressed from the seeds of the *Garcinia indica* tree. It is superbly soothing to the skin and regularly enjoyed in creams and lotions. Kokum butter will melt effortlessly at body temperature and is a terrific ingredient in lip-balm recipes.

Illipe butter (3): A creamy or white hard butter that is expeller-pressed from illipe nuts from the *Shorea stenoptera* tree. Illipe butter is an extraordinary conditioning ingredient and an ideal addition to lotions and creams that will be used to rejuvenate and refresh dehydrated and thirsty skin.

Mango butter (4): An off-white-coloured semi-hard butter that is expeller-pressed from the seed kernels of the mango tree (*Mangifera indica*). Mango butter has a substantial amount of both antioxidants and essential fatty acids. It is frequently used when preparing recipes for nourishing and soothing dry skin, as well as smoothing wrinkles. Mango butter is a superb ingredient in lotions, creams, body balms, lip balms and soaps.

Murumuru butter (5): An off-white-coloured hard butter that is expeller-pressed from the fruits of the *Astrocaryum murumuru* tree. It contains an extraordinary amount of essential fatty acids and is used in beauty products to rejuvenate and recondition dry and mature skin.

Shea butter (6): A creamy tan to pale-yellowish-coloured, velvety soft butter that is expeller-pressed from the nuts of the shea tree (*Vitellaria paradooxa*, formerly *Butyrospermum parkii*). It is one of the most commonly used butters in skincare recipes. It has a remarkable capacity to protect and soothe the skin, and is regularly used in lotions, creams, body balms and lip balms.

4

Using natural butters

This chart shows the shelf-life of the most commonly used natural butters if they are stored in an air-tight container in a cool, dark place. Also provided is the melting point of the butters, which is important to know if you are creating lip balms, which may warm up and potentially melt if kept close to the body.

Butter	Shelf-life	Melting Point
Cocoa butter	2 to 4 years	34°C (93°F)
Kokum butter	1 year	35°C (95°F)
Illipe butter	1 to 2 years	37°C (98°F)
Mango butter	2 years	37°C (98°F)
Murumuru butter	2 years	31°C (88°F)
Shea butter	1 year	28°C (82°F)

FIX IT

If your butter is too hard to incorporate into recipes, then melt it in a bain-marie over a low heat.

5

Make lemony whipped body butter

This body butter instantly penetrates the skin to provide long-lasting protection and moisture. See pages 94–95 and 112–113 for some more great face and body moisturiser recipes.

Ingredients:
- *55 g shea butter*
- *2 tablespoons cocoa butter*
- *2 tablespoons jojoba oil*
- *¼ teaspoon vitamin E oil*
- *40 drops lemon essential oil*

Makes approx. 110 g

1. Measure out the shea butter, cocoa butter and jojoba oil into a bain-marie set over a low heat.
2. Melt the butters and oil together, and remove from the heat.
3. Cool to room temperature and then place in the fridge for 20 minutes.

4. Using a hand-held mixer, with a whisk attachment, whisk for 10 minutes. Return the mixture to the fridge to chill for 5 minutes. Whisk again for 10 minutes and repeat the whisk/chill process until you have the consistency of whipped cream. Add the vitamin E oil and lemon essential oil, and whisk to combine.
5. Transfer the body butter to an air-tight container and store in a cool, dark place. Use within six months.

6

Tips and tricks for buying and storing butters

- Always store butters in air-tight containers.
- Keep butters in a cool, dark place.
- Use clean utensils when removing butters from their containers.
- Both unrefined shea and natural cocoa butter have a distinct aroma that will affect the fragrance of your finished product. You can use deodorised cocoa butter and refined shea butter if you wish to add a signature essential oil blend to your product.
- Only buy 100% natural butters that have not been hydrogenated or mixed with preservatives, fragrances or any other ingredient.

TRY IT

Massage pure shea butter into your tummy during pregnancy in order to keep your skin supple and help prevent stretch marks.

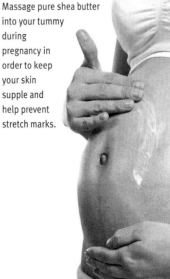

Clays

Mineral-rich clays are unearthed from natural quarries around the globe and have been utilised in skincare routines for hundreds of years to cleanse, tone and revitalise the face and body. Depending on the type of clay, they can be used to gently exfoliate skin or significantly absorb oils and tighten up pores. There is a perfect clay for every skin type.

Understanding the most popular types of clay

From the renowned bedrock quarries of France, the fertile Atlas Mountains of Morocco and the plentiful volcanic ash sediments in the United States, naturally occurring clays are rich in silica, magnesium, aluminium, calcium and other beneficial minerals, which make wonderful ingredients in skin-pampering beauty recipes.

Bentonite clay (1): Also known as sodium bentonite and sodium montmorillonite, this light grey, odourless and very fine clay has a high content of the minerals silica and aluminium. It is found in natural volcanic ash sediments in Montana and Wyoming, in the United States. Used in facial mud treatments, body powders, dry shampoos and scrubs.
Colour: Pale to light grey
Odour: Neutral
Price: Economically priced
Country of origin: Wyoming and Montana, United States
Notable mineral content: Silica, aluminium, iron, magnesium

French green clay (2): Also known as illite clay and sea clay, this light- to medium-green clay is mined from quarries that can be thousands of feet deep in France, China and the United States. French green clay is rich in silica, aluminium, calcium, iron and magnesium. A very fine-textured clay that is used to absorb oils and impurities from the face and body.
Colour: Light to medium green
Odour: Neutral
Price: Expensively priced
Country of origin: France, United States, China
Notable mineral content: Silica, aluminium, valcium, iron, magnesium, potassium

Fuller's Earth clay (3): Contains a high amount of silica, magnesium oxide and sapphire crystal. This off-white clay is the most popular clay for oily and congested skin. Very drying and oil absorbing.
Colour: Pale to off-white
Odour: Neutral
Price: Moderately priced
Country of origin: United States, Japan, Mexico
Notable mineral content: Silica, magnesium, iron oxide and sapphire crystal

Rhassoul clay (4): Also known as red Moroccan clay and red clay, this light greyish/pink clay comes from Morocco and is high in silica, magnesium, calcium and aluminium. This particular type of clay is used in spas throughout the world to pamper the skin.
Colour: Light grey to pinkish
Odour: Neutral
Price: Moderately to expensively priced
Country of origin: Morocco
Notable mineral content: Silica, aluminium, magnesium, calcium

White kaolin clay (5): Also known as white cosmetic clay and China clay, this pure white clay is used extensively in numerous beauty products, including soaps, face masks, natural deodorants and face and body scrubs and powders. High in kaolinite, silicon oxide and aluminium oxide.
Colour: Pure white
Odour: Neutral
Price: Economically priced
Country of origin: United Kingdom, United States, Germany, China, Australia
Notable mineral content: Kaolinite, silicon oxide, aluminium oxide

Choosing the best clay for your skin type

Natural clays have varying strengths for drawing out impurities and revitalising the skin. It is essential to use the most suitable clay for your particular skin type for the most beautiful results.

Skin Type	Type of Clay	Frequency of Use
Normal	Bentonite, French green, Fuller's Earth, rhassoul and white kaolin	Normal skin can tolerate the use of clay-containing formulas up to several times a week.
Oily & Blemish-Prone	Bentonite, French green, Fuller's Earth and rhassoul	The most drawing and oil-absorbing clay for oily skin is Fuller's Earth, which can be used up to twice a week.
Dry & Sensitive	Rhassoul and white kaolin	Dry and sensitive skin should only be exposed to clay-containing formulas once or twice a week and for a maximum of 15 minutes at a time.

How to make and use a basic clay face mask

This simple face mask does not call for any measuring or weighing out of ingredients, making it a quick and easy beauty fix. See pages 86–87 for some more great face mask recipes.

1. Measure out a heaping tablespoon of clay into a small bowl.
2. Mix in just enough warm water, herbal tea or floral water (hydrosol) to create a spreadable paste.
3. Apply a thick layer of clay to your cleansed face (avoiding the delicate eye area).
4. Allow the mask to dry and then rinse off with some warm water.

TRY IT

• When making a sensitive or dry skin face mask, add a bit of milk, honey or carrier oil to help pamper and moisturise your special skin.

• Add powdered rolled oats, crushed herbs and/or cocoa powder to face masks for a pampering experience.

Keep it pure

When purchasing clays, check that they are 'cosmetic grade' to ensure they are pure and do not contain high levels of lead or other harmful fillers.

FIX IT

Store clays in a sealed, moisture-proof container to maximise their shelf-life of more than 5 years to indefinitely.

Sugars and Salts

Sweet sugars and splendid salts can make delightful additives in many pampering skincare creations. Dip into your sugar bowl and craft a scrumptious and exfoliating sugar body scrub. Sprinkle some mineral-rich sea salts into a warm bath for a revitalising and stress-reducing spa treatment.

11

Choosing sugars

There are many different types of sugar used in beauty and skincare recipes. Sugars with the smallest crystal size are best in products for the face and for sensitive skin. Granulated white sugar is often used to make body, hand and foot scrubs.

Granulated white sugar (1): This particular type of sugar is what is commonly found at your supermarket and used in home baking. The crystal size in granulated white sugar is usually graded as 'fine' and is ideal for crafting body scrubs, hand polishes and foot exfoliators. Granulated white sugar should not be used in recipes for the face or used by those with delicate skin.

Coarse sugar (2): A white sugar that has a larger crystal size than granulated white sugar and is very coarse. This particular sugar should only be used in hand and foot treatments.

Caster sugar (3): Also known as baker's sugar, this has a smaller crystal size than granulated sugar and is good for making skincare recipes that will be used on the face and body.

Icing sugar (4): This is an ultra-fine sugar that has a very small crystal size. This is the best sugar choice when creating scrubs for the face and for those with delicate skin.

Demerara sugar (5): A brown sugar with a large crystal size that makes an excellent addition to body scrubs. It is light brown in colour and has a black treacle-like aroma that makes a sweet scrubbing treat. Avoid use on delicate skin and the face.

Evaporated cane juice (6): This sugar is light brown in colour and has a black treacle-like aroma, and is good used in body care recipes. Avoid using on delicate skin and the face.

Brown sugar (light and dark) (7): This is perfect for crafting body scrubs. If you are using this particular type of sugar on the face or on delicate skin, then use a gentle pressure when exfoliating to avoid irritation.

Muscovado sugar (8): A very dark brown sugar with coarse, large crystals that makes a great addition to foot and hand scrubs.

TRY IT

Got super-dirty hands from gardening? Mix 1 tablespoon of regular white sugar with 1 teaspoon of olive oil, 1 teaspoon of grated orange peel and 1 teaspoon of your favourite hand soap to form a scrub. Gently massage and scrub your hands to loosen up the dirt. Rinse with warm water.

FIX IT

If you don't have any caster sugar, simply add granulated sugar to a food-processor or blender, and process until you create a finer-grade sugar.

Choosing salts

There are several types of salt used in bath and body treatments. From relaxing and soothing bath soaks to skin-polishing salt scrubs, turn simple salts into extraordinary spa secrets.

Dead Sea salt (1): A pure white salt that is harvested from the Dead Sea in Israel. This particular salt has a very high mineral content and can be purchased either in coarse grade, which is perfect for bath salt recipes, or in fine grade, which is wonderful for body scrubs.

Himalayan bath salt (2): This red-, pink- and white-freckled salt is mined from within the Himalayan Mountains. When purchasing this particular type of salt, choose extra-coarse grade for making scented potpourri with essential oils, small to medium grade for bath salts and fine grade for making salt scrubs.

Sea salt (3): This is a white salt that is similar in texture to regular table salt and collected from evaporated ocean waters around the world. This economically priced salt is great for making salt scrubs and bath bombs. Choose extra small grade for most body scrub applications.

Grey sea salt (4): A light grey salt that is harvested from the island of Noirmoutier, near Brittany, in France. It is sold under the trademark Breton™ Grey Sea Salt and is moderately to expensively priced. This particular salt has a very high mineral content. Purchase fine or velvet grade for face and body scrubs and coarse grade for bath treatments.

Epsom salts (magnesium sulphate) (5): A white salt that is often used for creating relaxing bath soaks and calming foot baths. Many people find that soaking in an Epsom salt bath helps to soothe muscle cramps and ease stress. It can also be used to exfoliate the skin and help wash away foot odour. When purchasing this particular type of salt, choose medium to coarse grade for bath and foot soaks and extra fine for body scrubs.

Note: If you have diabetes and/or any health concerns, speak with your doctor before using Epsom salts.

Perfume pedi

Do you have stinky feet? Mix 135 g of Epsom salts and 15 g of dried lavender buds in a tub of warm water. Soak your feet for 30 minutes to freshen and revitalise your feet.

Soothing salt soak

Make a soothing bath salt soak by combining 275 g of Epsom salts, 1 tablespoon of freshly chopped rosemary and 1 tablespoon of freshly chopped spearmint in a closeable mesh bag and tossing it into a warm bath. Soak your stresses away!

TRY IT

If you have rough and dry elbows and knees, mix a handful of fine sea salt with a small amount of body soap and rub over your skin to exfoliate and smooth.

Waxes, Thickeners and Emulsifiers

When preparing skincare products such as lotions and conditioners, it is essential to include ingredients such as waxes, thickeners and/or emulsifiers in your formulations. Many of these unique ingredients, including beeswax, lanolin, lecithin and xanthan gum, are easy to find at your local natural market, whereas emulsifying wax, stearic acid and candelilla wax can all be purchased at specialist shops and online.

FIX IT

Purifying beeswax

The most economically priced beeswax is raw and unfiltered, and comes straight from the extracting room. It will always contain a significant amount of debris, which will need to be removed before you incorporate it into skincare formulas.

1. Add the beeswax to a saucepan of low-simmering water until it has melted.

2. Run a fine-mesh metal sieve with a handle through the hot water to remove floating debris.

3. Carefully filter the hot melted wax and water through several layers of ultra-fine muslin placed in a fine-mesh sieve to remove all the debris and pollen into another heat-safe bowl.

4. Allow the beeswax to solidify and harden. Simply remove the hardened beeswax disc from the water and pat dry. It is important to use a separate saucepan and bowl for this process.

15

Waxes

These are multifaceted mixtures of esters, fatty acids and alcohols. Waxes such as beeswax are hard substances and are impervious to moisture, which, in turn, makes them resistant to degradation. Waxes are frequently used to thicken recipes such as lip balms, salves and lotion bars. Waxes also offer beneficial and skin-protecting properties. Waxes have a high melting point of 49–102°C (120–215°F).

Beeswax (1): Excreted by worker honeybees, beeswax is comprised of a mixture of wax esters and fatty acids. Beeswax has a melting point of 60–68°C (140–155°F). It is soluble in oil and is essential when crafting salves and lotion bars. Beeswax can be purchased in sizable blocks, small 25-g bars and convenient pastilles. Unrefined beeswax is a natural yellow colour with a rich, intoxicating, honey-like aroma. It may also be purchased as refined white granules with no detectable aroma. Beeswax is perfect for use in lotions, creams, conditioners, balms, body butters, lip balms and salves.

Carnauba wax (2): This is an extremely hard wax exuded by Brazilian palm tree leaves.

This wax is comprised of fatty acids, wax esters and fatty alcohols. The wax has a light-yellow hue with a neutral aroma. It has a high melting point of 79–88°C (175–190°F). Perfect for making lip balms, salves, stiff creams and skin-barrier protective formulas.

Candelilla wax (3): This is a vegetable wax from the candelilla plant (*Euphorbia antisyphilitica*), which grows in Mexico. It has a pale yellow hue with no detectable aroma. Its melting point is 68–74°C (155–165°F). It is ideal for making lip balms, salves, body butters and skin-barrier protective formulas.

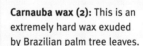

16

Thickeners

Thickeners are used to enhance the consistency and/or viscosity of beauty products. They are also used as humectants to help retain water on the skin.

Guar gum (1): This is a pale yellow powder from the seeds of the guar plant (*Cyamopsis tetragonoloba*). It is used as a thickener and viscosity enhancer in creams, lotions and conditioners. It is dissolved in the water phase of recipes in a concentration of 0.5 to 2%.

Gum Arabic (2): Also known as acacia gum, this water-soluble ingredient comes from the African acacia tree and is used in oil/water formulas, and also as an emulsifier and thickener. It can be found in resinous chunks or ground into a fine, white powder and has a neutral aroma. Choose the powder form where possible for ease of use. It is usually added to the water phase of recipes in a concentration of 1 to 10%. Great in cream and lotion formulas.

Xanthan gum (3): An odourless white powder, xanthum gum is excreted from the bacteria *Xanthomas campestris*. It is soluble in water and used to improve the viscosity and volume of lotions, creams, shampoos and cleansers. Xantham gum is normally used in recipes at a concentration of 0.5 to 2%.

17

Emulsifiers

These are used to unite water and oils together into a homogenous and stable concoction. Emulsifying wax is the most common type of emulsifier and is used when making lotions and creams.

Emulsifying wax NF (1): This is a plant-based emulsifier from the fatty acids of plant fats. It may be purchased in convenient white pastilles with no detectable aroma. It has a melting point of 52°C (125°F). Perfect when making lotions, creams and conditioners. Used in a typical concentration of 2 to 6% in most formulas.
Note: This particular ingredient can endure considerable processing and also contains polysorbates. This means that it cannot be categorised as '100% Natural', although it is a well-accepted additive in most natural recipes. 'NF' indicates that the ingredient meets the standards of the National Formulary.

Anhydrous lanolin (2): A yellow, waxy substance secreted by the sebaceous glands of sheep. Anhydrous lanolin is sold without containing any water. It enhances the viscosity of lotions, creams and lip balms, and acts as a mild emulsifier. Perfect when used in skin-barrier protective formulas. It is used in concentrations of 2 to 20%.

Lecithin (3): This is a natural lipid that is capable of binding water and oils together. Most commonly derived from soya beans and chestnut-brown in colour, it is a honey-like substance with a nutty aroma. It is used to emulsify lotions and creams. It is added to the oil phase of a lotion or cream recipe in a concentration of 0.5 to 5%. Liquid lecithin may be purchased in containers or in gel caps that may be squeezed out. It is best to purchase non-GMO soya-based lecithin where possible.

Stearic acid (4): These white flakes are derived from vegetable fats. This naturally occurring fatty acid makes a wonderful emulsifier and thickener for creams, lotions and shave creams. Used in the oil phase of a recipe in a concentration of 2 to 10%.

Liquid Carrier Oils

Natural carrier oils are extracted from both abundant and exotic botanical seeds, nuts, pits, kernels, beans and pulps. Carrier oils may be obtained from the biggest avocado pit and the teeniest blackberry seed. Loaded with nutritious vitamins, beneficial fatty acids, natural tocopherols (vitamin E) and an abundance of other superb skin-nurturing compounds, carrier oils are essential when formulating countless categories of skincare recipes.

18

Methods of extraction

The best techniques for naturally extracting oils from their sources are the cold-pressed method and the expeller-pressed method. Both of these time-honoured approaches preserve the complex, health-giving components of the oils. Steer clear of carrier oils that have been obtained using harsh extraction methods, which require intense heat, solvents, chemicals or extreme pressure. Organic versions of many carrier oils may be purchased, but they will probably cost significantly more.

The expeller-pressed method: The raw material that contains the wholesome oil is deposited into a portly, barrel-shaped apparatus, which boasts forceful, screw-like teeth that crush and masticate the raw material. Constant pressure causes the oil to separate from the raw material and seep out. Due to the powerful grinding and persistent pressure involved in this extraction method, the temperature of the oil can increase to 93°C (200°F).

The cold-pressed method: This is identical to the expeller-pressed method except for the use of a chilling device to ensure that the temperature of the oil does not go higher than 38–43°C (100–110°F). Due to the delicate character of certain carrier oils, it is very important that gentle heat is used during this extraction process in order to conserve the favourable properties of the oil.

◀ Natural carrier oils come in a variety of subtle hues, viscosities and aromas. Take their differences into consideration when using them in a formula as they will impart distinctiveness to the product.

Oil-refining methods

As soon as the carrier oil has been extracted, it is frequently refined to filter, deodorise, bleach or fractionate it. Sometimes, a combination of these treatments is used.

Unrefined: The raw oil is simply filtered through a screen or sieve to get rid of any unwanted crude material such as shell fragments, fibres or pulp. The natural colour, aroma and flavour of the raw oil are left intact.

Partially refined: The raw oil is filtered and may undergo one or more additional refining methods, including being deodorised to eliminate the aromatic and volatile compounds, naturally bleached with clays or activated charcoal and/or fractionated to ensure that the oil does not become cloudy or viscous when exposed to cold temperatures.

Fully refined: The raw oil is filtered, deodorised, naturally bleached and fractionated. It is also often flash-heated to make it shelf-stable. Fully refined oils are sourced when the formulator wants a carrier oil that will not impart any aroma or colour to a finished product. Fully refined carrier oils are not as beneficial to the skin because many of the curative compounds are removed during the refining process.

Sunflower-soft hands

If you experience dry, chapped hands while cooking in the kitchen, reach for some culinary-grade sunflower oil. Massage into your hands and continue cooking without concern.

Sourcing carrier oils

If you can't locate a specific carrier oil at your supermarket, then you may have better luck going to a health food shop or natural market to find a greater collection of natural oils.

Thrifty tips

Are luxurious oils breaking your beauty budget? Try these cash-savvy tips and keep your cupboard stocked without breaking the bank.

- Try mixing an expensively priced oil such as pomegranate oil with a moderately or economically priced oil, such as argan oil, sunflower oil or sesame oil, along with sweet almond oil.
- Choose small-sized bottles of oils to try out and use. Many suppliers will offer tester bottles.
- Kill two birds with one stone by looking for oils that are manufactured for both culinary and cosmetic use. For example, olive oil is as good for you externally, as it is internally.

TRY IT

- Apply a small amount of jojoba oil to split ends in your hair to smooth them.

- Massage a mixture of 4 drops of baobab oil and 1 drop of vitamin E oil into your nails and cuticles before bed for supple and healthy-looking nails.

▼ Olive oil is a very versatile and nourishing carrier oil for all skin types.

Profiles of key carrier oils

Carrier Oil Common and Botanical Name	Preferred Extraction Method	Characteristics and Pricing	Recommended Skin Types and Benefits	Storage and Shelf-life
Almond or sweet almond oil (*Prunus dulcis*)	Expeller-pressed and can be unrefined or partially refined	• Golden colour • Nutty aroma • Economically priced	**Skin types:** All **Benefits:** • Emollient, soothing and conditioning • Easily absorbed	Store in a cool, dark place for up to 2 years.
Apricot kernel oil (*Prunus armeniaca*)	Cold-pressed and unrefined	• Golden colour • Nutty aroma • Moderately priced	**Skin types:** Sensitive, mature and dry **Benefits:** • Easily absorbed • Protecting and soothing • High in oleic fatty acid	Store in a cool, dark place for up to 2 years.
Argan oil (*Argania spinosa*)	Cold-pressed and unrefined	• Golden colour • Subtle aroma • Expensively priced	**Skin types:** All **Benefits:** • High in vitamin E, antioxidants and fatty acids • Softening, hydrating and strengthening • Good for stretch mark and anti-wrinkle formulas	Store in a cool, dark place for up to 1 year.
Avocado oil (*Persea americana*)	Cold-pressed and unrefined	• Dark rich green colour • Avocado-like aroma • Economically priced	**Skin types:** All, but especially sensitive, dry and mature **Benefits:** • High in vitamins, amino acids and fatty acids • A rich oil that pampers the skin • Great for eczema and psoriasis formulas	Store in a fridge for up to 18 months.
Baobab oil (*Adansonia digitata*)	Cold-pressed and unrefined	• Golden yellow colour • Subtle nutty aroma • Expensively priced	**Skin types:** All, especially dry and mature **Benefits:** • Rich in vitamins • Absorbs quickly into the skin • Great for both skin and hair treatments	Store in a cool, dark place for up to 2 years.
Blackberry seed oil (*Rubus fruticosus*)	Cold-pressed and unrefined	• Light- to medium-green colour • Subtle nutty aroma • Expensively priced	**Skin types:** All **Benefits:** • High antioxidant protection • High in vitamin E • Great in lip balms, face creams and eye creams	Store in a cool, dark place for up to 1 year.
Blueberry seed oil (*Vaccinium corymbosum*)	Cold-pressed and unrefined	• Light green colour • Subtle aroma • Expensively priced	**Skin types:** All **Benefits:** • High antioxidant protection • High in vitamin E • Great in lip balms, face creams and eye creams	Store in a cool, dark place for up to 1 year.

Carrier Oil Common and Botanical Name	Preferred Extraction Method	Characteristics and Pricing	Recommended Skin Types and Benefits	Storage and Shelf-life
Camelina oil (*Camelina sativa*)	Expeller-pressed and unrefined	• Dark golden colour • Herbaceous aroma • Economically priced	**Skin types:** Normal, combination, sensitive, dry and mature **Benefits:** • High in antioxidants and vitamin E • Great in both skin and hair recipes • Soothing and hydrating	Store in a cool, dark place for up to 2 years.
Castor oil (*Ricinus communis*)	Expeller-pressed and refined	• Very light golden colour • Undetectable aroma • Economically priced	**Skin types:** All, especially irritated and sensitive **Benefits:** • Very protective and barrier-forming • Used in small amounts, along with other carrier oils, makes wonderful, protective and rejuvenating creams and lotions • Adds a wonderful texture to thick salves	Store in a cool, dark place for up to 2 years. **Note:** Do not ingest!
Coconut oil (*Cocos nucifera*)	Cold-pressed or expeller-pressed and unrefined or fully refined	• White colour • Strong coconut aroma (unrefined) to undetectable (fully refined) • Economically priced	**Skin types:** Normal, combination, sensitive, dry and mature **Benefits:** • Protective and may relieve irritated skin • Great in lip balms, lotions, creams and hair treatments	Store in a cool, dark place for up to 2 years.
Cranberry seed oil (*Vaccinium macrocarpon*)	Cold-pressed and unrefined	• Slight green colour • Subtle aroma • Expensively priced	**Skin types:** All **Benefits:** • High in omega fatty acids • High in antioxidants and vitamin E • Non-greasy and fast absorbing	Store in a cool, dark place for up to 1 year.
Evening primrose seed oil (*Oenothera biennis*)	Cold-pressed and partially refined	• Golden yellow colour • Subtle nutty aroma • Moderately priced	**Skin types:** Normal, sensitive, dry and mature **Benefits:** • Protective, soothing and rejuvenating • Great in formulas for dry and irritated skin	Store in a fridge for up to 1 year.
Grapeseed oil (*Vitis vinifera*)	Cold-pressed and unrefined or partially refined	• Light green to dark green colour • Subtle nutty aroma • Economically priced	**Skin types:** All, especially dry and sensitive **Benefits:** • Quickly absorbed • Adds a silky feel to lotions and creams • Good for eczema and psoriasis formulas	Store in a cool, dark place for up to 2 years.
Hazelnut oil (*Corylus avellana*)	Expeller-pressed and partially refined	• Clear colour • Subtle nutty aroma • Economically priced	**Skin types:** Normal, combination, oily and blemish-prone **Benefits:** • The best carrier oil for oily and blemish-prone skin due to its astringent quality	Store in a cool, dark place for up to 2 years.
Hemp seed oil (*Cannabis sativa*)	Cold-pressed and unrefined	• Dark green colour • Subtle nutty aroma • Economically priced	**Skin types:** All, especially dry and mature **Benefits:** • Easily absorbed into the skin • Rich in vitamins and fatty acids • Rejuvenating and protecting	Store in a fridge for up to 2 years.

Carrier Oil Common and Botanical Name	Preferred Extraction Method	Characteristics and Pricing	Recommended Skin Types and Benefits	Storage and Shelf-life
Jojoba oil (*Simmondsia chinensis*)	Cold-pressed and unrefined	• Golden colour • Subtle aroma • Expensively priced	**Skin types:** All **Benefits:** • Frequently used in beauty recipes for all skin-types so a good investment • Quickly absorbed since it is similar to our own sebum. May be used by itself to moisturise the skin and scalp	Store in a cool, dark place for up to 3 years.
Kukui nut oil (*Aleurites moluccans*)	Cold-pressed and fully refined	• Light golden colour • Subtle nutty aroma • Economically priced	**Skin types:** All, especially sensitive and irritated skin **Benefits:** • Absorbs very quickly into the skin and is highly moisturising and protective	Store in a cool, dark place for up to 2 years. **Note:** Heat-sensitive, should be added to formulas at under 38°C (100°F).
Macadamia nut oil (*Macadamia integrifolia*)	Expeller-pressed and unrefined	• Light golden colour • Sweet nutty aroma • Economically priced	**Skin types:** Normal, sensitive, dry and mature **Benefits:** • Wonderful carrier oil when making recipes to help nourish and regenerate the skin • Easily absorbed and perfect for use in formulas meant to protect, heal and rejuvenate	Store in a cool, dark place for up to 2 years.
Meadowfoam seed oil (*Limnanthes alba*)	Expeller-pressed and fully refined	• Light golden colour • Undetectable aroma • Moderately priced	**Skin types:** All **Benefits:** • Great for skin and hair recipes • Absorbs easily, leaving no greasy residue	Store in a fridge for up to 2 years.
Neem oil (*Azadirachta indica*)	Cold-pressed and unrefined	• Dark colour • Very strong garlic and peanut-like aroma. Usually only used in therapeutic recipes for skin and scalp • Moderately priced	**Skin types:** All **Benefits:** • Usually mixed with another carrier oil so will go far • Great in formulas meant for blemishes and skin problems	Store in a cool, dark place for up to 2 years.
Olive oil (*Olea europaea*)	Cold-pressed and unrefined	• Deep green colour • Rich olive-like aroma • Economically priced	**Skin types:** Normal, sensitive, dry and mature **Benefits:** • A wonderful, all-round carrier oil to be used in all beauty recipes calling for a nourishing and protecting oil	Store in a cool, dark place for up to 2 years.
Pomegranate seed oil (*Punica granatum*)	Cold-pressed and unrefined	• Light amber colour • Subtle fruity aroma • Expensively priced	**Skin types:** All **Benefits:** • A very luxurious and nutritious oil with high antioxidant properties	Store in a cool, dark place for up to 2 years.

Carrier Oil Common and Botanical Name	Preferred Extraction Method	Characteristics and Pricing	Recommended Skin Types and Benefits	Storage and Shelf-life
Pumpkin seed oil (*Cucurbita pepo L.*)	Cold-pressed and unrefined	• Rich green colour • Subtle nutty aroma • Moderately priced	**Skin types:** All **Benefits:** • Nourishing • High in vitamins and fatty acids	Store in a fridge for up to 2 years.
Rosehip seed oil (*Rosa canina*)	Cold-pressed and partially refined	• Rich amber colour • Strong aroma • Expensively priced	**Skin types:** Normal, sensitive, dry and mature **Benefits:** • Nourishing and protective • High in essential fatty acids • Great in recipes to rejuvenate the skin and treatments for wrinkles and mature skin	Store in a fridge for up to 2 years.
Sea buckthorn oil (*Hippophae rhamnoides*)	Cold-pressed and unrefined	• Dark amber to reddish colour • Strong aroma • Expensively priced	**Skin types:** All, especially sensitive and dry **Benefits:** • Super-high in essential fatty acids, vitamin E, and carotenes • Perfect when included in formulas for eczema, wrinkles and problem skin	Store in a cool, dark place for up to 2 years. **Note:** Must be highly diluted or will stain skin.
Sesame oil (*Sesamum indicum*)	Expeller-pressed and unrefined	• Dark golden colour • Strong nutty aroma • Moderately priced	**Skin types:** All **Benefits:** • Easily absorbed • Imparts a silky feel to creams and lotions • Great for a massage oil and used in hair treatments	Store in a cool, dark place for up to 2 years.
Sunflower oil (*Helianthus annuus*)	Expeller-pressed and fully refined	• Very light yellow colour • Subtle aroma • Economically priced	**Skin types:** All **Benefits:** • Easily absorbed • High in vitamins and oleic fatty acid • Perfect when included in formulas for dry, parched and mature skin • Good for both skin and hair recipes	Store in a cool, dark place for up to 2 years.
Tamanu oil (*Calophyllum inophyllum*)	Cold-pressed and unrefined	• Dark green colour • Strong and heavy aroma • Expensively priced	**Skin types:** All **Benefits:** • Supports skin healing and is wonderful in formulas to address wrinkles, eczema, stretch marks and mature skin	Store in a cool, dark place for up to 2 years.
Vitamin E oil (*Tocopherol*)	Unrefined mixed tocopherols from GMO sources are best	• Slight amber colour • Heavy, strong aroma • Expensively priced	**Skin types:** All **Benefits:** • A wonderful addition to all skincare products for its superb antioxidant protection • Helps prolong the shelf-life of creams, lotions, salves and other beauty products by protecting the oils from rancidity	Store in a cool, dark place for up to 2 years. **Note:** Not to be used undiluted on the skin
Wheat germ oil (*Triticum vulgare*)	Cold-pressed and unrefined	• Dark amber to light brown colour • Heavy, strong aroma • Expensively priced	**Skin types:** Normal, sensitive, dry and mature **Benefits:** • High in vitamins • Great for dry and rough skin and wrinkle formulas	Store in a fridge for up to 2 years. **Note:** Avoid if you have a wheat or gluten allergy.

Herbal-Infused Carrier Oils

Many natural beauty recipes depend on herbal-infused oils as a base ingredient. By infusing an oil with the healing and curative properties of herbs, you can create natural beauty formulations that are gentle, protecting, repairing and healing. Discover which herbs are the best choices and the unique ways of infusing them into your favourite oil.

◀ Fresh thyme

23

Make a dried-herb-infused oil – folk method

The folk method involves no measuring or weighing of ingredients, so it's a simple way to create an infused oil.
1. Place the dried herb into a small sterile Mason jar with lid.
2. Pour enough carrier oil into the jar to cover the dried herb by at least 2.5 cm. Stir the mixture to completely saturate the herb with the oil. Add more oil if necessary. Cap the jar tightly and place in a paper bag.
3. Place the jar in a warm spot, such as a window. Shake the jar daily for six weeks.
4. Strain the oil out of the herb by using muslin and squeezing the oil out into a clean sterile jar. Tightly cap and store in a cool dark place for up to six months.

24

Make a dried-herb-infused oil – quick method

When you can't wait six weeks for your carrier oil to infuse, use the quick method for same-day results.
1. Place the dried herb into a slow cooker or bain-marie. Pour in enough carrier oil to cover the dried herb by at least 2.5 cm. Gently heat the herb and oil over the lowest heat of 38–54°C (100–130°F) for 8 hours, stirring frequently. Allow the herbs to cool to room temperature.
2. Strain the oil out of the herb by using muslin and squeezing the oil out into a clean sterile jar. Tightly cap and store in a cool dark place for up to six months.

25

Make the perfect healing camper's salve

Ingredients:

- *60 ml herbal-infused olive oil*
- *2 heaping tablespoons beeswax*
- *10 drops lavender essential oil*
- *10 drops tea tree essential oil*
- *10 drops myrrh essential oil*

Makes approx. 85 g

This salve is perfect for a summer first aid kit. First infuse your olive oil with burdock root, calendula flowers, chamomile flowers, chickweed, goldenseal, nettles, Oregon grape root, plantain, thyme and yarrow. The salve is thickened with beeswax, and lavender, tea tree and myrrh essential oils. These herbs have antibacterial, anti-inflammatory, antiseptic, anti-toxic and general skin-soothing properties.

1. Grind the herbs to form a powder and create a herbal-infused oil using one of the methods described above.
2. Once ready, heat the herbal-infused oil and beeswax until melted, then mix in the other ingredients.
3. Pour into your chosen container and allow to cool.

FIX IT

To keep your herbal-infused oil from turning rancid, make sure you use dried herbs, such as you would find in your local supermarket, and choose an oil with a long shelf life, such as olive oil or jojoba oil.

Choosing beneficial herbs for oil infusing

Botanical Name	Common Name	Infused Oil Qualities
Achillea millefolium	**Yarrow**	Good base oil for making recipes to soothe dry, damaged skin.
Althaea officinalis	**Marshmallow root**	Good base oil for making recipes for dry, mature skin.
Arctium lappa	**Burdock root**	Good base oil for hair treatments.
Calendula officinalis	**Calendula flowers**	Good for dry and damaged skin, and for making soothing salves and creams.
Hydrastis canadensis	**Goldenseal**	Good base oil when making recipes for acne and blemish-prone skin.
Lavandula x intermedia	**Lavender flowers**	Good base oil for making all types of beauty products. Smells wonderful and is great for massage oils.
Mahonia aquifolium	**Oregon grape root**	Good base oil when making recipes for oily skin, acne and blemish-prone skin.
Matricaria recutita	**Chamomile flowers**	Good base oil for making all types of beauty products. Smells wonderful and is great for massage oils.
Melissa officinalis	**Lemon balm**	Good for general skin-soothing recipes. Good for making lip balms to soothe cold sores.
Mentha piperita	**Peppermint**	Great base oil when formulating foot-care products.
Plantago major	**Plantaina**	Good base oil for all skin types.
Rosmarinus officinalis	**Rosemary**	Good base oil for hair treatments.
Sambucus nigra	**Elder flower**	Good base oil for making all types of beauty products. Smells wonderful and is great for massage oils.
Stellaria media	**Chickweed**	Great for healing and protecting skin that is dry and damaged. Great for scalp treatments.
Symphytum officinale	**Comfrey root and leaf**	Good base oil when formulating recipes for sunburn, eczema, insect bites, stretch marks and dry skin. **Note:** Not to be used on dirty wounds or open skin.
Thymus vulgaris	**Thyme**	Good base oil when making recipes for acne and blemish-prone skin.

Essential Oils

The art of aromatherapy relies on the use of highly complex and concentrated essential oils, which are extracted from various plant sources such as leaves, flowers, berries, petals and fruit. These volatile compounds are the 'soul' or essence of the plant from which they are extracted. When you stop to smell a beautiful rose, it is the essential oil in the petal cells that produces the intoxicating and delightful aroma. Essential oils are frequently used in a natural beauty applications. They can be used to formulate perfumes and fragrances (see pages 128–135); support and promote healthy, radiant skin; and can even be diffused into the air to stimulate your mind or calm your nerves.

Buying essential oils for skincare and beauty applications

It is important to purchase and use only genuine and therapeutic, high-quality essential oils from trustworthy and well-regarded sources. Unfortunately, there are companies that deal in adulterated and/or synthetic essential oils, which are ineffective and can also be dangerous. Here are a few tips on choosing high-quality authentic essential oils:

- Choose 'Certified Organic' or 'Wild-harvested' essential oils whenever possible.
- Store essential oils in amber or dark-coloured glass bottles, with childproof caps if you have small children at home.
- Choose small bottles (15 ml) as many essential oils are best used within one to two years of purchasing.
- Buy pure undiluted essential oils that have not been diluted with carrier oils or other liquids.
- Purchase from companies that are reputable and are available to answer any questions you may have.
- Many places that sell essential oils will have 'tester' bottles for you to try before you purchase. This is a great way to become familiar with the aroma of these essential oils.

Choosing essential oils according to price

A significant factor that will determine the price of an essential oil is the amount of plant material it takes to produce the oil. For example, Bulgarian rose (*Rosa damascena*) essential oil is one of the most expensive essential oils on the market because it takes approximately 1,000,000 freshly picked rose flowers to steam-distill 1 litre of essential oil. In contrast, sweet orange (*Citrus sinensis*) essential oil is very economically priced due to a readily available supply of fresh orange peels that yield a substantial amount of essential oil when cold-pressed.

ECONOMICALLY PRICED ESSENTIAL OILS (£5 OR LESS)	MODERATELY PRICED ESSENTIAL OILS (£6–15)	EXPENSIVELY PRICED ESSENTIAL OILS (£16 OR MORE)
Anise seed	Bay laurel	Angelica root
Basil	Bergamot	Cardamom
Benzoin resin oil	Black pepper	Carrot seed
Bitter orange	Cinnamon bark	Chamomile, blue
Cedarwood, Atlas	Clary sage	Chamomile, Moroccan
Cedarwood, Virginia	Cypress	Chamomile, Roman
Cinnamon leaf	Fennel	Cistus
Citronella	Frankincense	Davana
Clove bud	Galbanum	Douglas fir
Copaiba balsam	Geranium, rose	Grand fir
Coriander seed	Ginger	Helichrysum
Elemi	Hyssop	Hops flower
Eucalyptus	Juniper berry	Jasmine absolute
Eucalyptus, lemon	Lavender	Lemon balm
Fir needle	Mandarin	Manuka
Grapefruit	Marjoram, sweet	Melissa
Lavender 40/42	Nutmeg	Myrrh
Lavender, spike	Opopanax/Sweet myrrh	Myrtle
Lemon	Patchouli	Neroli/Orange flower
Lemongrass	Sage, common	Oak moss absolute
Lime peel	Thyme, red	Rose absolute
Litsea cubeba	Vetiver	Rose essential oil, Bulgarian
Niaouli	Ylang ylang	Sandalwood, Australian
Orange, bitter		Spikenard
Orange, sweet		St. John's wort
Palmarosa		Vanilla absolute
Peppermint		Yarrow, blue
Peru balsam		
Petitgrain		
Pine, Scotch		
Ravensara		
Rosemary		
Rosewood		
Savory, winter		
Spearmint		
Spruce		
Tangerine		
Tarragon		
Tea tree		

• *Please note: In this chart the common name of the essential oil is given, while the prices are for 15 ml bottles at 2014 prices. Please see the full Essential Oils chart on pages 32–37 for the botanical names of all the oils.*

FIX IT

• If an essential oil seems too inexpensive, there is a good chance that it is not a worthy essential oil. You should only purchase high-quality oils and the price should be within true market-price ranges (see left).

• You can try expensively priced oils without breaking the bank by purchasing small bottles of 4 ml or less.

Some helpful definitions

Essential oil: A concentrated liquid that is extracted from plant parts. Essential oils contain volatile aroma compounds that are not soluble in water. The molecules are very tiny and can evaporate quickly, which allows them to be absorbed into our bodies by inhalation and/or by applying them to the skin. Essential oils can be extracted by steam-distillation or expression.

Absolutes: These are solvent extracts of plant material. Jasmine absolute is an example in which a hydrocarbon solvent is used to extract the aroma molecules from the flowers. Absolutes are mostly used to make perfumes and fragrant applications.

CO_2 extracts: Also known as supercritical CO_2 extracts, these are very similar to steam-distilled essential oils. CO_2 extraction is a modern way of extracting volatile oils using supercritical pressurised carbon dioxide.

Citrus essential oils: Most citrus fruit essential oils are obtained by cold-pressing the fruit peels. It is important to choose citrus essential oils that have been certified organic whenever possible.

28

Got milk?

Essential oils do not mix well with water. So, add your essential oil to a tablespoon of double cream or milk before adding to bath water.

29

Breathe yourself happy

Breathing in essential oils can promote a balancing physiological effect. Try placing a few drops of essential oil onto a tissue and breathing in the vapours. Try a few drops of lavender for a calming effect, a few drops of basil for an uplifting and energising effect, a few drops of rose geranium to help wake you up when you get tired in the middle of the day, or a few drops of peppermint to help ease the symptoms of a headache.

31

Essential oil weight and measurement conversions

Depending on the viscosity of the essential oil, there are approximately 20 to 30 drops of essential oil per millilitre (ml). Essential oils are usually measured in drops from either the orifice reducer in the bottle or via a pipette. The measurements and weights given in the chart below are approximate.

Millilitres	Teaspoons	Fluid ounces	Drops from a pipette or orifice reducer
3.75 ml	$^3/_4$ teaspoon	$^1/_8$ fl oz	75 to 112 drops
7.5 ml	1 teaspoon	$^1/_4$ fl oz	150 to 224 drops
15 ml	3 teaspoons	$^1/_2$ fl oz	300 to 448 drops
30 ml	6 teaspoons	1 fl oz	600 to 896 drops

Remember: 1 millilitre of essential oil = 20 to 30 drops

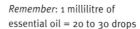

30

Avoid irritation

When blending with essential oils for prolonged periods of time, wear a pair of disposable gloves to avoid direct contact with your hands, which could lead to irritation.

Tips on using essential oils safely

Essential oils have been used safely for a very long time. Here are some important guidelines to follow when using essential oils:

- Keep all essential oils out of the reach of children and pets. Store essential oils in childproof containers when necessary.
- Never apply an essential oil undiluted. Follow the recommendations for proper dilution.
- Do not use essential oils on babies or children, unless you are working with a qualified practitioner.
- Always perform a patch test using 2% diluted essential oil in a carrier oil applied to your inner arm and wait for 24 hours to check for redness, irritation or any adverse reactions. Do not use if an adverse reaction occurs. Seek medical help if necessary (see page 11).
- Never use essential oils internally, unless you are working with a qualified practitioner.
- Avoid getting essential oils in your eyes, ears, nose or mouth, or in contact with other sensitive mucus membranes.
- If you have sensitive skin, cardiovascular or kidney problems, epilepsy or any other medical conditions, do not use essential oils unless you have been advised by your health care practitioner that it is safe.
- If you are pregnant or breastfeeding, seek advice from your healthcare professional before using essential oils.
- If you are taking any medications for any medical problem, do not use essential oils unless you have been advised by your health care practitioner that it is safe.
- Do not go into direct sunlight for at least eight hours after applying a product that contains a photosensitising oil.
- If an essential oil gets into your eyes, immediately flush them with cold milk or vegetable oil to dilute the essential oil. Seek medical attention as soon as possible.
- If an undiluted essential oil gets onto your skin, use a cream or vegetable oil to dilute and then wash with soap and water.
- If an essential oil is ingested, notify your doctor immediately and seek appropriate medical treatment.

FIX IT

- Diffusing essential oils into the air is a wonderful way of experiencing their therapeutic benefits. Follow the directions on your nebuliser or diffuser for the best results (see page 135).

- If your potpourri has grown stale, sprinkle it with a few drops of your favourite essential oil to liven it up and make your living space smell amazing.

'Essential' essential oils

Whether you are just starting out with essential oils and would like to create a beginner's kit or you have an extensive knowledge of aromatherapy and would like to complement your current collection, here is my favourite ensemble of essential oils. My criteria for deciding on this selection included incorporating a balanced array of essential oils that are generally considered safe, are traditionally used for their skin-restorative benefits and properties, are aromatically delightful and are offered 'Certified Organic'.

If your budget only allows you to purchase one or two essential oils at the beginning, choose lavender and/or frankincense. Both are moderately priced and are wonderful additions in most skincare products. One or both may offer benefits for blemishes, dry skin, scars, wrinkles sunburns and more.

Lavender (*Lavandula angustifolia*)

Frankincense (*Boswellia carterii*)

Helichrysum (*Helichrysum italicum*)

Carrot seed (*Daucus carota*)

Bulgarian rose (*Rosa damascena*)

Tea tree (*Melaleuca alternifolia*)

Blue chamomile (*Matricaria recutita*)

Sandalwood (*Santalum spicatum*)

Geranium (*Pelargonium graveolens*)

Palmarosa (*Cymbopogon martinii*)

(34)

Essential oil profiles

For each essential oil, this chart provides information on the method of extraction, the plant part used, properties and safety considerations. The information is for educational purposes only – it is not intended to treat, cure, prevent or diagnose any disease or condition. Nor is it intended to prescribe in any way. This information has not been evaluated by the Medicines and Healthcare Products Regulatory Agency (MHRA). Please note that all of the essential oils listed over the following 6 pages should be used highly diluted.

Essential Oil	Method of Extraction & Plant Part Used	Potential Therapeutic Properties & Benefits	Special Safety Cautions
Angelica root (*Angelica archangelica*)	Steam-distilled roots	Antibacterial, antifungal, antispasmodic, stimulant, tonic	• Avoid while pregnant. • May be photosensitising.
Anise seed (*Pimpinella anisum*)	Steam-distilled seeds	Analgesic, antiseptic	• Avoid while pregnant or breast-feeding. • May cause skin irritation.
Basil (*Ocimum basilicum*)	Steam-distilled flowering plant	Antibacterial, antiseptic, antispasmodic, stimulant, tonic	• Avoid in epilepsy or while pregnant. • May cause skin irritation.
Bay laurel (*Laurus nobilis*)	Steam-distilled leaves and twigs	Analgesic, anesthetic, antibacterial, antifungal, antimicrobial, antiseptic, sedative	• Avoid while pregnant. • May cause skin irritation.
Benzoin resin oil (*Styrax tonkinensis*)	Ethanol extraction of the resin	Anti-inflammatory, antioxidant, antiseptic, astringent, deodorant, sedative, styptic	• Avoid while pregnant. • Not for internal use.
Bergamot (*Citrus bergamia*)	Cold-expressed fruit peel	Analgesic, antibacterial, antidepressant, antiseptic, antispasmodic, astringent, deodorant, sedative, stimulant, tonic	• Avoid while pregnant. • May be photosensitising.
Bitter orange (*Citrus aurantium*)	Cold-expressed fruit peel	Anti-inflammatory, antiseptic, antispasmodic, astringent, bactericidal, deodorant, fungicidal, stimulant	• May cause skin irritation. • May be photosensitising.
Black pepper (*Piper nigrum*)	Steam-distilled dried fruit	Analgesic, antibacterial, antimicrobial, antiseptic, antispasmodic, aphrodisiac, stimulant, tonic	• Avoid with homeopathics, in kidney or liver disease, and while pregnant. • May cause skin irritation.
Cardamom (*Elettaria cardamomum*)	Steam-distilled fruit	Antiseptic, antispasmodic, nerve tonic, stimulant	
Carrot seed (*Daucus carota*)	Steam-distilled seeds	Antiseptic, stimulant, tonic	• Avoid while pregnant. • May be photosensitising.
Cedarwood, Atlas (*Cedrus atlantica*)	Steam-distilled wood and sawdust	Antifungal, antiseptic, aphrodisiac, astringent, regenerative, sedative, tonic	• Avoid while pregnant.
Cedarwood, Virginia (*Juniperus virginiana*)	Steam-distilled wood	Antiseptic, antispasmodic, astringent, circulatory stimulant, sedative	• Avoid while pregnant.

Essential Oil	Method of Extraction & Plant Part Used	Potential Therapeutic Properties & Benefits	Special Safety Cautions
Chamomile, blue (*Matricaria recutita*)	Steam-distilled flowers	Analgesic, anti-inflammatory, antispasmodic, bactericidal, fungicidal, nerve sedative	
Chamomile, Moroccan (*Ormenis mixta*)	Steam-distilled flowers	Antispasmodic, sedative	
Chamomile, Roman (*Anthemis nobilis*)	Steam-distilled flowers	Analgesic, antibacterial, anti-inflammatory, antimicrobial, antiseptic, antispasmodic, sedative, tonic	
Cinnamon bark (*Cinnamomum zeylanicum*)	Steam-distilled dried inner bark	Analgesic, antibacterial, antifungal, anti-inflammatory, antimicrobial, antiseptic, antispasmodic, aphrodisiac, astringent, stimulant	• Avoid while pregnant and in liver or kidney disease. • Do not use on skin unless highly diluted. • Not for internal use.
Cinnamon leaf (*Cinnamomum zeylanicum*)	Steam-distilled leaves	Analgesic, antibacterial, anti-inflammatory, antiseptic, antispasmodic, stimulant	• Avoid while pregnant. • May cause skin irritation.
Cistus (*Cistus ladaniferus*)	Steam-distilled aerial part	Antimicrobial, antiseptic, astringent, tonic	• Avoid while pregnant.
Citronella (*Cymbopogon winterianus*)	Steam-distilled grass	Analgesic, antiseptic, astringent, deodorant	• Avoid while pregnant. • May cause skin irritation.
Clary sage (*Salvia sclarea*)	Steam-distilled leaves and flowers	Antibacterial, antiseptic, antispasmodic, aphrodisiac, astringent, deodorant, euphoric, sedative	• Avoid while pregnant.
Clove bud (*Syzygium aromaticum*)	Steam-distilled flower buds	Analgesic, anti-aging, antibacterial, antifungal, anti-inflammatory, antimicrobial, antispasmodic, antioxidant, antiseptic, antiviral, stimulant	• Avoid while pregnant and in liver and kidney conditions. • May cause skin irritation.
Copaiba balsam (*Copaifera officinalis*)	Steam-distilled crude balsam	Antibacterial, anti-inflammatory, disinfectant, stimulant	• Avoid while pregnant. • May cause skin irritation.
Coriander seed (*Coriandrum sativum*)	Steam-distilled seeds	Analgesic, antibacterial, anti-rheumatic, antispasmodic, aphrodisiac, fungicidal, revitalising, stimulant, tonic	• Avoid while pregnant.
Cypress (*Cupressus sempervirens*)	Steam-distilled needles and twigs	Antibacterial, anti-inflammatory, antiseptic, antispasmodic, astringent, deodorant, sedative, tonic	• Avoid while pregnant.
Davana (*Artemisia pallens*)	Steam-distilled leaves and flowers	Antiseptic, antiviral, aphrodisiac, disinfectant, sedative	• Avoid while pregnant.
Elemi (*Canarium luzonicum*)	Steam-distilled gum	Analgesic, antiseptic, antiviral, fungicidal, regulatory, stimulant, tonic	• Avoid while pregnant.
Eucalyptus (*Eucalyptus globulus*)	Steam-distilled leaves and twigs	Analgesic, antibacterial, antifungal, antiseptic, antispasmodic, antiviral, deodorant, stimulant	• Avoid while pregnant and with homeopathics. • May cause skin irritation.

Essential Oil	Method of Extraction & Plant Part Used	Potential Therapeutic Properties & Benefits	Special Safety Cautions
Eucalyptus, lemon (*Eucalyptus citriodora*)	Steam-distilled leaves and twigs	Antiseptic, antiviral, bactericidal, calmative, deodorant, fungicidal	• Avoid while pregnant and with homeopathics.
Fir needle (*Abies balsamea*)	Steam-distilled needles	Analgesic, antiseptic, antitussive, astringent, deodorant, stimulant, tonic	• Avoid while pregnant.
Frankincense (*Boswellia carterii*)	Steam-distilled resin	Analgesic, antifungal, anti-inflammatory, antioxidant, antiseptic, astringent, sedative, tonic	
Galbanum (*Ferula galbaniflua*)	Steam-distilled resin	Analgesic, anti-inflammatory, antimicrobial, antiseptic, antispasmodic, hypotensive, restorative, tonic	
Geranium, rose (*Pelargonium graveolens*)	Steam-distilled flowers and leaves	Analgesic, antibacterial, antidepressant, anti-inflammatory, antiseptic, astringent, deodorant, regenerative, sedative, styptic, tonic, vasoconstrictor	• Avoid while pregnant. • May cause skin irritation.
Ginger (*Zingiber officinale*)	Steam-distilled root	Analgesic, antibacterial, anti-inflammatory, antioxidant, antiseptic, antispasmodic, aphrodisiac, astringent, stimulant, tonic	• May cause skin irritation. • May be phototoxic.
Grapefruit (*Citrus paradisi*)	Cold-expressed fruit peel	Antibacterial, antidepressant, antiseptic, astringent, restorative, stimulant, tonic	• May cause skin irritation. • May be phototoxic.
Helichrysum (*Helichrysum italicum*)	Steam-distilled flowers	Antibacterial, anti-inflammatory, antimicrobial, antioxidant, antispasmodic, astringent, stimulant	• Avoid while pregnant.
Hops flower (*Humulus lupulus*)	Steam-distilled flowers	Antimicrobial, antiseptic, antispasmodic, astringent, bactericidal, sedative	• Avoid in depression and while pregnant. • May cause skin irritation.
Hyssop (*Hyssopus officinalis*)	Steam-distilled flowering plant	Antibacterial, antiseptic, antispasmodic, antiviral, astringent, sedative, tonic	• Avoid while pregnant.
Jasmine absolute (*Jasminum grandiflorum*)	Ethyl alcohol solvent extracted from flowers	Analgesic, anti-inflammatory, aphrodisiac, tonic	• Avoid while pregnant.
Juniper berry (*Juniperus communis*)	Steam-distilled berries	Analgesic, antimicrobial, antiseptic, antispasmodic, astringent, sedative	• Avoid in kidney or liver disease and while pregnant.
Lavender (*Lavandula angustifolia*)	Steam-distilled flowering tops	Analgesic, antibacterial, anti-inflammatory, antimicrobial, antiseptic, antispasmodic, aromatic, deodorant, sedative, stimulant	
Lavender, spike (*Lavandula latifolia*)	Steam-distilled flowering tops	Analgesic, antibacterial, anti-inflammatory, antimicrobial, antiseptic, antispasmodic, aromatic, deodorant, stimulant	• Avoid while pregnant.
Lemon (*Citrus limon*)	Cold-expressed fruit peel	Antibacterial, antifungal, anti-inflammatory, antimicrobial, antiseptic, antispasmodic, astringent, sedative	• Avoid while pregnant. • May be photosensitising.

Essential Oil	Method of Extraction & Plant Part Used	Potential Therapeutic Properties & Benefits	Special Safety Cautions
Lemon balm (*Melissa officinalis*)	Steam-distilled flowering plant	Antibacterial, antihistaminic, anti-inflammatory, antiseptic, antispasmodic, antiviral, bactericidal, sedative, tonic	• Avoid while pregnant.
Lemongrass (*Cymbopogon flexuosus*)	Steam-distilled grass	Analgesic, antifungal, anti-inflammatory, antimicrobial, antioxidant, antiseptic, antiviral, astringent, bactericidal, deodorant, fungicidal, sedative, tonic	• Avoid while pregnant. • May be photosensitising.
Lime peel (*Citrus aurantifolia*)	Cold-expressed fruit peel	Antibacterial, antiseptic, antispasmodic, antiviral, astringent, bactericidal, deodorant, restorative, tonic	• Avoid while pregnant. • May be photosensitising.
Litsea cubeba (*Litsea cubeba*)	Steam-distilled fruit	Antibiotic, anti-infectious, anti-inflammatory, antiseptic, deodorant, sedative, stimulant	• Avoid while pregnant.
Mandarin (*Citrus reticulata*)	Cold-expressed fruit peel	Antiseptic, antispasmodic, hypnotic, lymphatic stimulant, sedative, tonic	• Avoid while pregnant.
Manuka (*Leptospermum scoparium*)	Steam-distilled leaves and twigs	Analgesic, anesthetic, antibacterial, antifungal, anti-inflammatory, antimicrobial, antiseptic, antiviral, deodorant, sedative	• Avoid while pregnant.
Marjoram, sweet (*Marjorana hortensis*)	Steam-distilled flowering plant	Analgesic, antioxidant, antiseptic, antispasmodic, antiviral, sedative, tonic	• Avoid while pregnant.
Melissa (*Melissa officinalis*)	See Lemon balm		
Myrrh (*Commiphora myrrha*)	Steam-distilled gum	Antifungal, anti-inflammatory, antimicrobial, antiseptic, antispasmodic, antiviral, astringent, fungicidal, sedative, tonic	• Avoid while pregnant.
Myrtle (*Myrtus communis*)	Steam-distilled leaves and twigs	Antiseptic, astringent, bactericidal, sedative, tonic	• Avoid while pregnant.
Neroli/Orange flower (*Citrus aurantium*)	Steam-distilled flowers	Antibacterial, anti-inflammatory, antiseptic, antispasmodic, aphrodisiac, fungicidal, sedative, tonic	• Avoid while pregnant.
Niaouli (*Melaleuca viridiflora*)	Steam-distilled leaves and twigs	Analgesic, antiseptic, antispasmodic, bactericidal, stimulant	• Avoid while pregnant.
Nutmeg (*Myristica fragrans*)	Steam-distilled seeds	Analgesic, antioxidant, antiseptic, antispasmodic, aphrodisiac, stimulant, tonic	• Avoid while pregnant.
Oak moss absolute (*Evernia prunastri*)	Solvent extraction of oak tree lichen	Antiseptic, demulcent, fixative	• Avoid while pregnant.
Opopanax/Sweet myrrh (*Commiphora holtziana*)	Steam-distilled resin	Antifungal, anti-inflammatory, antimicrobial, antiseptic, antispasmodic, sedative	• Avoid while pregnant. • May be photosensitising.
Orange, bitter (*Citrus aurantium*)	Cold-expressed fruit peel	Anti-inflammatory, antiseptic, antispasmodic, astringent, bactericidal, deodorant, fungicidal, stimulant	• Avoid while pregnant.

Essential Oil	Method of Extraction & Plant Part Used	Potential Therapeutic Properties & Benefits	Special Safety Cautions
Orange, sweet (*Citrus sinensis*)	Cold-expressed fruit peel	Anti-inflammatory, antiseptic, antispasmodic, bactericidal, fungicidal, stimulant, tonic	• May be photosensitising.
Palmarosa (*Cymbopogon martinii*)	Steam-distilled grass	Antibacterial, antifungal, antiseptic, antiviral, stimulant, tonic	• Avoid while pregnant. • May be photosensitising.
Patchouli (*Pogostemon cablin*)	Steam-distilled leaves	Antibacterial, anti-inflammatory, antimicrobial, antiseptic, antiviral, bactericidal, deodorant, stimulant, tonic	
Peppermint (*Mentha piperita*)	Steam-distilled flowering plant	Analgesic, antibacterial, anti-inflammatory, antifungal, antimicrobial, antiseptic, antispasmodic, astringent, sedative, stimulant, vasoconstrictor	• Avoid while pregnant.
Peru balsam (*Myroxylon balsamum*)	Steam-distilled crude balsam	Anti-inflammatory, antiseptic, stimulant	• Avoid while pregnant.
Petitgrain (*Citrus aurantium*)	Steam-distilled leaves and twigs	Antiseptic, antispasmodic, deodorant, stimulant, tonic	• Avoid while pregnant.
Ravensara (*Agathophyllum aromatica*)	Steam-distilled leaves	Analgesic, antibacterial, anti-infectious, antiseptic, antiviral, stimulant	• Avoid while pregnant.
Rose absolute (*Rosa damascena*)	Alcohol solvent extraction of flower petals	Antiviral, aphrodisiac, astringent, sedative, tonic	• Avoid while pregnant.
Rose essential oil, Bulgarian (*Rosa damascena*)	Steam-distilled flower petals	Analgesic, antibacterial, antimicrobial, antiseptic, antiviral, aphrodisiac, astringent, bactericidal, deodorant, disinfectant, sedative, tonic	• Avoid while pregnant.
Rosemary (*Rosmarinus officinalis*)	Steam-distilled flowering tops	Analgesic, antibacterial, antioxidant, antiseptic, antispasmodic, aphrodisiac, astringent, fungicidal, restorative, stimulant, tonic	• Avoid while pregnant and with hypertension.
Sage, common (*Salvia officinalis*)	Steam-distilled leaves	Antibacterial, anti-inflammatory, antimicrobial, antioxidant, antiseptic, antispasmodic, astringent, tonic	• Avoid while pregnant.
Sandalwood, Australian (*Santalum spicatum*)	Steam-distilled roots and heartwood	Antiseptic, antispasmodic, aphrodisiac, astringent, bactericidal, emollient, fungicidal, sedative, tonic	
Spearmint (*Mentha spicata*)	Steam-distilled flowering plant	Analgesic, anesthetic, antibacterial, anti-inflammatory, antiseptic, antispasmodic, astringent, stimulant, tonic	
Spikenard (*Nardostachus jatamansi*)	Steam-distilled roots	Antibiotic, antifungal, anti-infectious, anti-inflammatory, antiseptic, bactericidal, deodorant, fungicidal, sedative, tonic	

Essential Oil	Method of Extraction & Plant Part Used	Potential Therapeutic Properties & Benefits	Special Safety Cautions
Spruce (*Tsuga canadensis*)	Steam-distilled needles	Antimicrobial, antiseptic, astringent, tonic	• Avoid while pregnant.
Tangerine (*Citrus reticulata*)	Cold-expressed fruit peel	Antimicrobial, antiseptic, antispasmodic, hypnotic, stimulant, tonic	• May be photosensitising.
Tarragon (*Artemisia dracunculus*)	Steam-distilled flowering plant	Antiseptic, antispasmodic, hypnotic, stimulant	• Avoid while pregnant.
Tea tree (*Melaleuca alternifolia*)	Steam-distilled leaves and twigs	Analgesic, antibacterial, antifungal, anti-inflammatory, antimicrobial, antiseptic, antiviral, deodorant, fungicidal	
Thyme, red (*Thymus zygis*)	Steam-distilled flowering plant	Analgesic, antibacterial, antifungal, anti-inflammatory, antimicrobial, antioxidant, antiseptic, antispasmodic, antiviral, bactericidal, cell proliferant, deodorant, stimulant, tonic	• Avoid while pregnant. • May irritate skin.
Vanilla absolute (*Vanilla planifolia*)	Ethyl alcohol solvent-extracted bean and/or pod	Used in perfume making.	
Vetiver (*Vetiveria zizanoides*)	Steam-distilled roots	Analgesic, antibacterial, antifungal, anti-inflammatory, antimicrobial, antioxidant, antiseptic, antispasmodic, cell proliferant, sedative, stimulant, tonic	
Yarrow, blue (*Achillea millefolium*)	Steam-distilled flowers	Anti-inflammatory, antibacterial, antifungal, antipyretic, antiseptic, antispasmodic, astringent, stimulant, tonic	• Avoid while pregnant.
Ylang ylang (*Cananga odorata*)	Steam-distilled flowers	Used in perfume making.	• Avoid while pregnant.

35

Essential oil application and dilution

Never apply essential oils undiluted to the skin. Refer to this dilution chart for safe and effective results. If you have sensitive skin, try a dilution ratio of 0.5% (4 to 5 drops per 30 ml of carrier oil) to avoid irritation.

Application	Dilution Ratio	Amount of Essential Oil Added to Carrier Oil
Massage oil	2.5%	20 to 25 drops per 25 ml of oil
Bath treatment	5%	45 to 50 drops per 25 ml of oil-based carrier added to water
Pedicure soak	5%	45 to 50 drops per 25 ml of oil-based carrier added to water
Manicure soak	3%	27 to 30 drops per 25 ml of oil-based carrier added to water
Facial steam	1.5%	30 to 40 drops added to bowl of water
Facial mask	2.5%	20 to 25 drops per 25 ml of carrier oil
Facial treatment oil (leave on)	2%	18 to 20 drops per 25 ml of carrier oil
Facial cleansing oil (rinse/wipe off formulas)	3%	30 to 35 drops added to carrier oil
Hair oil treatment	2.5%	20 to 25 drops per 25 ml of carrier oil
Body perfume	5 to 10%	45 to 100 drops added to carrier oil or alcohol carrier
Body lotion	2.5%	20 to 25 drops per 25 ml of carrier oil

Floral Waters

Floral waters, or hydrosols, are the aromatic waters produced by steam-distilling a variety of plant and flower materials in copper stills. The term 'hydrosol' is Latin (*hydro* and *sol*) for 'water solution'. Floral waters are made during the production of essential oils and contain all of the enchanting plant essences but are much milder and usually contain less than 1 per cent essential oil, making them a wonderful ingredient for pampering skincare recipes.

Eight ways to use floral waters

There are many different ways in which you can use floral waters, some of which are outlined below:

1 Use as facial and body toners, either alone or in combination with other ingredients.

2 Use as the water phase in lotion and cream formulations to provide a beneficial ingredient or as an aromatic addition to the recipe.

3 Use as compresses to soothe tired and puffy eyes.

4 Use as the water phase when mixing a clay mask.

5 Use as an addition to herbal hair rinses and shampoos.

6 Use as the base for aromatic body sprays and perfumes.

7 Use to make your own bath bombs and fizzes.

8 Apply to cotton pads and use to remove makeup.

Understanding the distillation process

This illustration of a conventional still explains how steam passes through plant material and tranports its volatile constituents out through the top tunnel of the distillation chamber. The distillate then condenses to form the hydrosol (floral water) and essential oil. The essential oil accumulates on the hydrosol's surface, and the two are drained and separated into individual containers.

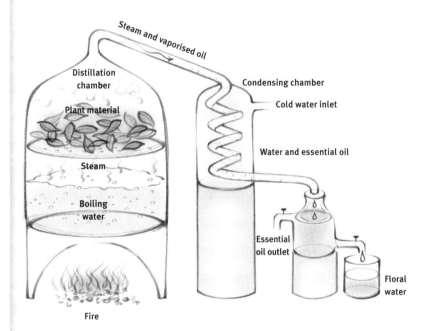

Free from fingers

Never use your fingers or unsanitary utensils to touch hydrosols. Use clean pipettes or glass dropper bottles to measure out floral waters.

TRY IT

Mix equal parts of neroli and rose floral waters in a spray bottle for a light and breezy body spray.

39

Picking the right floral water

Basil hydrosol: Distilled from the leaves of *Ocimum basilicum*. Try on blemish-prone skin as it has antibacterial properties.

Calendula hydrosol: Distilled from the flowers of *Calendula officinalis*. Try as a toner for blemish-prone or irritated skin.

Catnip hydrosol: Distilled from the flowers and leaves of *Nepeta cataria*. Try on camping or hiking trips as a mild insect repellant.

Chamomile hydrosol: Distilled from the flowers of *Matricaria recutita*. Great in all skincare recipes. Try misting over sunburned skin to soothe and reduce inflammation.

Clary sage hydrosol: Distilled from *Salvia sclarea*. Try using as an astringent toner on puffy or oily skin.

Cucumber hydrosol: Distilled from the fruit of *Cucumis sativus*. Try it chilled as an after-sun body mist to cool and hydrate sun-parched skin.

Lavender hydrosol: Distilled from the flowers of *Lavandula angustifolia*. Try as the water phase in your favourite cream or lotion recipe.

Lemon balm hydrosol: Distilled from the leaves of *Melissa officinalis*. Try as the water phase in massage creams to relax and uplift your spirits.

Lemon verbena hydrosol: Distilled from the leaves of *Aloysia citriodora*. Try in recipes meant for acne-prone skin.

Lime hydrosol: Distilled from the fruit of *Citrus latifolia*. Try as a refreshing chilled toner on parched skin.

Neroli (orange blossom) hydrosol: Distilled from the flowers of *Citrus aurantium*. Try this as an all-over-body mist after a warm bath to soothe and revitalise.

Peppermint hydrosol: Distilled from the aerial parts of *Mentha piperita*. Try misting this all over your tired feet for instant relief.

Rose geranium hydrosol: Distilled from the aerial portion of *Pelargonium capitatum*. Try using as the water phase in a clay mask for oily skin.

Rose hydrosol: Distilled from the petals of *Rosa damascena*. Try as a hair-refreshing mist during the day.

Rosemary hydrosol: Distilled from the leaves of *Rosmarinus officinalis*. Try in creams and lotions meant for rejuvenating the skin.

Tulsi (holy basil) hydrosol: Distilled from the aerial parts of *Ocimum tenuiflorum*. Try adding to lotions for blemish-prone skin.

Witch hazel hydrosol: Distilled from the twigs and bark of *Hamamelis virginiana*. Try as an astringent toner for oily and blemish-prone skin.

Yarrow hydrosol: Distilled from *Achillea millefolium*. Try as an addition to rich creams meant to alleviate damaged-skin conditions such as eczema.

40

Storing floral waters

Floral waters are very delicate due to their significant water content and the absence of any preservatives. They do not last as long as essential oils. For this reason, special consideration needs to be given to storing them correctly. Under appropriate storage, you can anticipate your floral water remaining at its best for up to one year.

- Store floral waters in tightly capped, dark-amber-coloured glass bottles to best protect them from harmful UVA and UVB rays.
- Store your floral waters in a cool, dark place such as a cupboard or pantry.
- If you notice a change in your floral water, i.e. cloudiness, mould, floating bits and pieces, bacterial blooms or a change in aroma, dispose of the floral water, as it has lost its sanitary integrity.
- If your hydrosols seem to have a short shelf life, try storing them in spray bottles to avoid opening the bottle and exposing them to oxygen and germs.

1
2
3

Herbs

If you are familiar with plant botany and are able to identify plants and herbs accurately, then you may enjoy wild-harvesting your favourite herbs. However, if you feel happier simply purchasing them from reputable suppliers, there are a number of ways to go about this.

 41

Where to buy fresh herbs

You can buy fresh herbs from a number of different sources, as follows:

Supermarkets: These sell bunches of fresh herbs, including basil, mint, dill, lemongrass, thyme, rosemary and other culinary herbs. Fresh herbs are often organic.

Farmers' markets: Providing a wonderful resource for fresh herbs, these markets usually start in spring and run through the summer. Fresh herbs are often sold in bulk, so you can purchase as much as you need.

Local farmers: Farmers who grow specific herbs are usually delighted to sell them to the public. If you need fresh lavender, for example, look for local lavender farmers.

Online retailers: There are online companies that can express-ship fresh herbs to you. They usually require a minimum purchase and the shipping can be expensive.

 42

Where to buy dried herbs

You can obtain dried herbs from various sources, as follows:

Supermarkets: The spice aisle at your local supermarket typically carries a wide variety of dried herbs. Many well-stocked shops will also offer bulk amounts of dried herbs.

Nutrition shops: These often stock dried herbs and you can sometimes even find herbs in bulk at such shops.

Herb shops: Typically operated by knowledgeable herbalists who can suggest specific herbs and offer detailed information for each herb they sell.

Online retailers: Those that specialise in dried herbs are a good choice, as the herbs are usually very fresh and the sales team will be knowledgeable about the various herbs on offer.

▲ Fresh thyme (1), bergamot (2), lovage (3), marjoram (4), parsley (5) and rosemary (6) can be wonderful additions to your home-made beauty creations.

43

Grow your own

Local garden and hardware shops typically offer herb seed packets that you can sprinkle in the garden to grow your own fresh herbs. Online herb seed companies are a great way to find a wide variety of plant seeds.

44

Using herbs in topical skincare applications

Herbs can be very soothing and healing when applied topically to the skin. Topical herbal preparations include:
• Herbal-infused lotions
• Herbal-infused creams
• Herbal-infused baths
• Herbal-infused salves and balms
• Herbal-infused oils
• Floral waters (hydrosols)
• Herbal compresses
There are many books available that explain in detail how to make the numerous topical applications. See page 138 for a recommendation.

◀ Dried herbs such as alkanet root (1), neem leaf (2), dried thyme (3) and lemon balm (4) should be stored in a cool, dark place.

45

Tips for buying and storing herbs

The best herbs are those that are 'Certified Organic' or 'Wild-harvested'. You should avoid herbs that have been treated with harmful chemicals. Here are some useful tips on buying and storing herbs:
• When buying fresh herbs, only purchase an amount that you can use up quickly. Unless you plan to dry fresh herbs, purchasing small amounts at a time is the best approach.
• Select fresh herbs that have a strong, herbaceous aroma and look freshly cut and healthy.
• Store fresh herbs in a cool, dry location such as the fridge. Dried herbs should be stored in air-tight containers in a cool, dark, dry place.
• Fresh herbs should generally be used up within a week or two. Once the fresh herbs begin to wilt or develop dark spots, they are past their best.
• Dried flowers, leaves and roots can last up to 12 months if they are stored properly. Dried seeds and bark can last up to 30 months if stored correctly.
• Always label herbs, especially dried herbs. Once dried herbs are cut and sifted, many flowers and leaves can look very similar to one another.
• Learn about any contraindications for various herbs. A herb may be contraindicated for various reasons. For example, it's recommended that ginger and yarrow should be avoided during pregnancy. If you have any medical issues, are taking medication or are pregnant, make sure the herbs you will be using are safe for your individual situation.

2 Tools and Supplies

One of the main advantages of creating kitchen-crafted natural beauty products is that you will probably already have in your kitchen many of the utensils and equipment needed to prepare the recipes in this book. Some people prefer to have a separate set of tools appointed just for fashioning beauty products. In this chapter you can work out exactly what tools you need and learn some clever tips and tricks on discovering quality equipment at reasonable prices.

Essential Kitchen Equipment

There are two basic categories of utensils and equipment needed for handcrafted beauty products. You will use certain utensils and equipment to prepare the recipes and need specific containers to store your beauty products in. Here we look at your 'making' tools.

46

Essential tools for product preparation

The great thing about making beauty products in your kitchen is that you will already have many of the tools you need to get started right away.

Measuring spoons: A good-quality set of measuring spoons made from a dishwasher-safe material, such as stainless steel, melamine or plastic, is a good choice for measuring small amounts of raw ingredients. Measuring spoons come in 'standard-measure' and 'odd-measure' sizes and often show milliletres for measuring liquid ingredients as well. Look for the following sizes: $\frac{1}{4}$ teaspoon (1.25 ml), $\frac{1}{2}$ teaspoon (2.5 ml), 1 teaspoon (5 ml), $\frac{1}{2}$ tablespoon (7.5 ml) and 1 tablespoon (15 ml) measurements. Measuring spoons with 'dash', 'smidgen' and 'pinch' measurements are available too. A 'dash' is $\frac{1}{8}$ teaspoon. A 'smidgen' is $\frac{1}{16}$ teaspoon. A 'pinch' is $\frac{1}{32}$ teaspoon.

Measuring jugs: You will need a set of measuring jugs for measuring liquid ingredients. Look for measuring jugs made from tempered glass that are marked with easy-to-read quantity indicators. Open handles are important for safe gripping and pouring. Your liquid glass measuring jugs should be microwaveable, as well as oven- and dishwasher-safe. They should have quantities shown in milliletres, and many will show fluid ounces and pints as well. They are available in a wide range of capacities, from 100 ml to more than 2 l. Having at least two 250-ml glass measuring jugs is a good starting point for your collection.

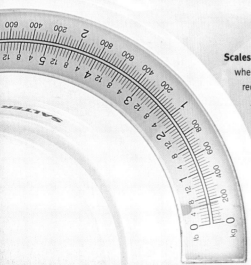

Scales: Scales are very important and convenient when exact ingredient measurements are required. Most scales can weigh in both grams and in ounces. A good-quality scale will have a tare feature, which enables you to weigh multiple ingredients into a single container. Battery-operated digital scales are the most accurate.

Mortars and pestles: A mortar and pestle is an important tool for crushing seeds and herbs, mashing fruit and combining essential oils into dry ingredients such as clays. You should buy a mortar and pestle made from granite or nonporous porcelain. It is not advisable to purchase the ones made from wood, as they can absorb essential oils.

Food grinders: A small spice and nut grinder is a wonderful way to grind whole spices, nuts, rolled oats, dried flowers, sugars and salts. You can find both electric models and ones that are hand-cranked.

Cooking utensils: This category includes whisks, stirring spoons, slotted spoons, skimmers, spatulas, ladles and other task-specific utensils used to prepare beauty recipes. You can find cooking utensils made from silicone, plastic, ceramic and nonreactive stainless steel. They need to be dishwasher-safe and heat-resistant.

Graters: These are the perfect tool for grating beeswax. Choose a dishwasher-safe grater that has razor-sharp, stainless-steel teeth and a comfortable grip handle.

Herb tools: Many recipes call for the addition of fresh and/or dried herbs. Some optional herb tools that may prove convenient include: herb-snipping scissors, herb choppers and herb mills. If you want to dry your own herbs from the garden, a food dehydrator may be useful.

Thermometers: These can be purchased as either digital or regular models. Thermometers are useful for determining the temperature of liquids.

Cutting boards: Commonly made from wood, plastic, marble and even stone, you can use a cutting board to chop ingredients for recipes. Many people keep a designated and separate cutting board just for their beauty recipes.

Wooden chopsticks: These are wonderful and economically priced tools that work well as stirring sticks. They are a good choice when mixing small amounts.

Silicone moulds: These come in a variety of shapes and sizes and are useful when making solid lotion bars and fizzy bath bombs. They are made from pliable silicone for easy product removal.

Pipettes and droppers: Made of either glass or plastic, these are useful when measuring small amounts of essential oils into liquid ingredients. The most common sizes are 1 ml and 3 ml.

Saucepans and bain-maries: Stainless steel is a good choice when choosing a saucepan or bain-marie. You will find many recipes calling for a bain-marie to melt waxes and oils together. You will also enjoy using a bain-marie to soften stiff butters such as shea butter and cocoa butter for easy blending.

Mixing bowls: You can never have too many mixing bowls in the kitchen. A good choice is either stainless steel or tempered glass, which are suitable for both hot and cold ingredients, as well as for use with a hand-held mixer. Having at least one of each of the following or similar sizes is a good start for your collection: small mixing bowl (18-cm diameter/1.5-l capacity); medium mixing bowl (20-cm diameter/3-l capacity); and large mixing bowl (25-cm diameter/5-l capacity).

Tea kettles: Great for boiling water to make herbal infusions.

Timers: Useful when timing herbal infusions and decoctions.

Funnels and sieves: Funnels are the perfect tool for transferring your finished liquid products into narrow-mouthed storage containers such as bottles and jars. Fine-mesh sieves are good at straining liquid from herbs.

Hand-held mixer, blender, food-processor and immersion blenders: Mixing creams and lotions is easy with a hand-held mixer, blender and immersion blender. A food-processor comes in handy when mixing facial masks, and salt and sugar scrubs. A food-processor can also be used to finely grind rolled oats, salts and sugars.

Safety glasses: These provide protection for your eyes. They are useful when mixing ingredients and keeping essential oils and other ingredients from splashing or spraying into your eyes.

Disposable gloves: These can protect your hands from essential oils and other ingredients.

Aprons: Protect your clothing from splashes, spills and stains.

A useful beginner's kit

- 250-ml- and 500-ml-capacity glass measuring jugs
- Digital scales
- Set of measuring spoons with capacities of $\frac{1}{8}$, $\frac{1}{4}$, $\frac{1}{2}$ and 1 teaspoon and 1 tablespoon
- 2 medium-sized mixing bowls
- 2 small or medium saucepans
- Whisk
- Spatula
- Stirring spoons
- Sieve
- Funnel
- Safety goggles
- Hand-held mixer

FIX IT

- Shop at discount shops, charity shops and even car boot sales for quality kitchen equipment at low prices.

- Choose dishwasher-safe utensils and equipment that can be sanitised in order to keep your beauty recipes pure and safe.

Safety and Sanitation

The recipes in this book do not use preservatives, and many of them have a short shelf-life. Natural skincare products are extremely susceptible to bacteria and germs. By starting with a sanitised workspace, working cleanly and storing your products in sanitised containers, you will improve shelf-life and lessen the chance that anything will taint your product.

48

Tips on cleanliness and storage

Start sanitary: You must sanitise your workspace surfaces, equipment, utensils, hands and storage containers. In other words, everything that will come into contact with your ingredients will need to be sanitised. One of the best ways to sanitise your workspace is with 70% alcohol. You can spray it on your countertops and other hard surfaces, spray it on your utensils and even mist it over your hands. Alcohol is highly flammable, so attention and care is required not to use it near an open flame. Objects sprayed with alcohol will need to air-dry before being used. If your dishwasher has a 'sanitise' setting, you may use it to wash your containers and utensils. Another sanitising method is the use of bleach. Ordinarily, one tablespoon of bleach is added to 3.5 litres of water. You need to soak your storage containers and utensils for 20 minutes, rinse with hot water and allow to air-dry. Follow the directions on the bottle of bleach carefully for further detailed instructions for sanitising surfaces.

Stay sanitary: During the course of creating beauty products, it is imperative that your workspace and hands stay clean. One idea is to have a bottle of alcohol-based hand sanitiser nearby to clean your hands. Between uses, keep your utensils and equipment on hygienic kitchen paper instead of placing them on countertops where they may come in contact with germs and bacteria. Do not leave your windows open or fans on where mould and fungus can circulate in the air and come to rest on your product.

Store sanitary: Your storage containers should be sanitised before you transfer your finished products to them. It is also essential not to introduce any bacteria, germs, fungus or moulds to the product while you are using it. You should never use your fingers to get your product out of its container. Every time, use a sanitised spoon or utensil to retrieve the product. Avoid getting water into your products, because even a tiny amount can harbour germs and ruin them. Keep your finished products in a cool, dark place. Most creams, lotions and other water-based products should be stored in a fridge and used up quickly. You may not always recognise that a beauty product has become contaminated with germs, since most germs are microscopic. However, should a product change in colour, texture or consistency, or if it develops a different odour, the odds are that it is no longer safe to use and must be thrown out.

Note: The recipes in this book are meant for your own personal enjoyment and for friendly gift-giving. They are not meant for sale to the public. Companies that sell beauty and skincare products have them tested by a professional laboratory to ensure they are sterile and not contaminated with germs.

Eco-Friendly Containers and Packaging

Most people who appreciate natural skincare also welcome environmentally friendly packaging. Also known as natural, sustainable packaging and green packaging, it is easy to find eco-friendly containers and packaging options for your beauty creations. Not only are nearly all of these packaging options recyclable, they are made from recycled materials too. Eco-friendly containers are available in a wide variety of sizes, shapes, materials and colours.

▲ Compostable tea bags

49

Choosing the best storage containers

Glass bottles: These are an eco-friendly option for storing essential oils, toners, cleansers and other liquid products.

Most glass bottles are fully recyclable and are often manufactured from recycled content. They come in different sizes, ranging from a 2 ml capacity up to a 1-litre capacity. You can find them in amber- and cobalt-blue-coloured glass, which are designed to withstand some level of UV contamination. You can also choose clear glass when you want to show the product inside. There are several types of closure for glass bottles: standard caps, droppers, atomisers, pumps, corks, sauce tops and swing tops.

Glass jars: Perfect for storing creams, salves, body butters, balms, thick lotions, sugar scrubs, salt scrubs and other recipes that require a tight-sealing lid. Just like glass bottles, you can find amber, cobalt blue and clear options. You can also find a variety of sizes, from less than 20 ml to 250 ml. You normally have a choice of either a plastic or metal lid. Glass jars can be found in a variety of fun shapes too. There are round, faceted, hexagonal, oval and Victorian square jars.

Glass vials: These are perfect for storing serums, perfumes and small amounts of liquid products. You can find perfume vials in various sizes, with roller-ball applicators or droppers. Most small-sized glass vials come with orifice reducers. You can find glass vials in amber- and cobalt-blue-coloured glass, as well as clear and green glass. Sizes range from 1 ml up to 25 ml.

Tin containers: These versatile tins have a base and a closure top. They are perfect for lip balms, salves, lotions, creams and solid lotion bars. You can find them in a variety of sizes, from 10-ml up to 50-ml sizes. They are available with a variety of tin closures such as slip covers, square, hinge, screw, twist and clear tops.

Recyclable plastic bottles: Manufactured from opaque plastic, these are good for lotions, shampoos, body oils, toners and other beauty products. The best choices are those bottles made from recyclable HDPE. They come in a variety of sizes, both small and large. You can find closures in flip caps, pumps, sprays and foamers. Some speciality bottle companies offer plastic bottles in amber, blue, clear, black and green colours.

Eco-friendly paper packaging: These use recycled fibres to manufacture recyclable packaging. You can find paper-based product packaging for body powders, lip balms, bath salts and even fragrance containers.

Bags: These are ideal for loose herbs, bath teas and other dry ingredients. You can find cello bags, which are clear, vegetable cellulose bags, cotton muslin bags, burlap bags and compostable tea bags.

▶ There is a wealth of eco-friendly packaging available – look for glass, tin and fabric options for storing ingredients and finished products. Recyclable HDPE plastic bottles and containers are a great option and can be found in any size or with any closure that you could need.

Creating Labels and Gift-Giving

An especially enjoyable aspect of handcrafting natural beauty creations is crafting labels, tags and wrappings for them. The sky truly is the limit on just how imaginative and original you can be when fashioning perfect accessories for homespun products. You may want a modest, handwritten label on recycled paper or opt for an assortment of ribbons, flourishes, bows and twine to make it especially distinctive.

 50

How to label beauty products

You should always label natural beauty products, because you don't want to have to guess what is in an unlabelled jar. The information on the label may include the following:

- **Name of product:** For example: *Thick & Rich Almond-Carrot Cleanser*
- **Ingredients:** It is common practice to list the ingredients in order of the greatest down to the smallest quantity. It's also a good idea to list the ingredients by both their common and botanical name. For example, sweet almond oil would be listed on the label as 'Almond, sweet *(Prunus dulcis)* oil'.
- **Directions for use:** A short statement on how to use the product should be included. For example, if the product is a hand cream, you could write: 'Massage a generous amount of hand cream into hands as needed'.
- **Warning and caution statements:** You may want to include caution statements on your label such as 'Keep out of reach of children and animals', 'For external use only' or 'Avoid the eye area'.

- **Best-used-by date:** Since natural beauty products do not contain preservatives, they often have short shelf-lives. You may want to include such notes as: 'Does not contain preservatives. Keep refrigerated and use within two weeks'.

If you are keeping the product for yourself, you may decide to use a very simple label that merely states the name of the product. If the product is intended as a gift, you should include as much information as possible on the label. If you are going to market and sell your handcrafted products to the public, it's important to follow the government-mandated labelling requirements for the country in which you will be selling them. Many countries have very strict rules and regulations for the labelling information required when introducing cosmetics and skincare into international commerce.

◀ Here is an example of a properly labelled beauty product that can be used for either personal use or for gift-giving.

MOISTURISING EYE SERUM
DIRECTIONS FOR USE: Apply a small amount to the lower eye area and massage in gently. Use twice daily.
INGREDIENTS: Pomegranate (Punica granatum) seed oil, Jojoba (Simmondsia chinensis) seed oil, Helichrysum (Helichrysum italicum) essential oil
CAUTION: For external topical use only
BEST USED BY: Use within 6 months of opening

▼ There are countless imaginative ways to package your handmade beauty creations. Ribbon, tissue paper, organza, fresh and dried herbs and flowers can all provide that extra personal touch.

53

Packaging ideas

When presenting your labelled product as a gift, you may wish to make it extra-special by placing it in a decorative box or container or by wrapping it in some special paper tied with twine and ribbons. Choosing repurposed and recycled containers is a wonderful eco-friendly option. There are many 'tree-free' packaging options that rely on sustainable plant fibres such as cotton, palm leaf and bamboo in the construction of the packaging.

Some wonderful supplies that can help you create the perfect eco-friendly package for your special gift include:
- Tissue paper
- Twisted-paper-handle brown kraft bags
- Sheer organdy bags with satin drawstrings
- Natural jute cloth bags
- Natural cotton muslin bags
- Natural linen bags
- Natural burlap bags
- Various sized jewellers' boxes
- Gable boxes
- Pillow boxes
- Crinkle-cut paper shred
- Shredded aspen wood
- Hemp twine cord
- Paper raffia
- Organza mesh ribbon
- Cellulose acetate ribbon
- Natural burlap ribbon
- Bakers' twine
- Cotton curling ribbon
- Biodegradable bags

51

Stickers, clipart and stamping ideas

If you want to create beauty-product labels on a computer, there is a large assortment of computer software to help you get creative. Take advantage of the huge quantity of clipart and elaborate fonts that are available for purchase and free of charge. Generally, office-supply shops stock an ample variety of labels in various textures, colours and sizes.

If you would prefer a more handcrafted look, then head to your local craft shop and choose some stamps and ink to create your labels.

52

Thoughtful gift ideas

- Garnish wrapped packages with some sprigs of rosemary, lavender bundles, cinnamon sticks or dried roses for an extra-special herbal touch.
- Make a beautiful gift basket by arranging several handcrafted products in a terracotta flowerpot – the perfect gift for an avid gardener.
- If you like to sew, save larger fabric scraps for wrapping beauty gifts.
- Wrap your product in a new cotton tea towel and tie it with bakers' twine.
- Wrap a brand-new pair of lightweight socks around a jar of foot salve and secure with ribbon or twine with a decorative tag that instructs the recipient to apply the foot salve and wear the socks in bed.

3 Naturally Lustrous Hair

Caring for your gorgeous tresses and lovely locks is very simple when it is done naturally. Your hair is happier when you steer clear of synthetic chemicals, hair-stripping sulphates and stifling silicones, and instead choose to nurture your magnificent mane with beneficial botanical ingredients. Delight in this lush mix of plant-based hair-care formulas to cleanse, condition, treat and nourish.

Natural Hair Treatments

If you treat your hair with care and attention, it will be both beautiful and healthy. The best thing you can do for your hair is to care for it naturally and avoid conventional products that contain harsh chemicals, petroleum products, sulphates, synthetic dyes and other potentially harmful ingredients. Instead, choose to care for your hair with plant-based ingredients that will strengthen, smooth and moisturise.

Combating chlorine

- Chlorinated water can wreak havoc with your hair. Consider installing a water filter to your shower head at home in order to remove harsh chemicals before they reach your hair and skin.
- Apply a generous amount of hair oil or serum to your hair and scalp before swimming to protect them from harsh pool chemicals and chlorine. Reapply as needed.

Choosing between natural and conventional commercial hair-care products

Natural products: Natural hair care and styling products are often formulated with the botanical ingredients listed below. These can moisturise the hair, smooth the cuticles, increase the hair's natural shine and softness, guard against breakage, increase manageability and strength and protect the hair from free radicals, as well as help hair appear fuller, thicker and more healthy!

- Natural plant oils
- Natural plant extracts
- Natural essential oils
- Natural herbs
- Gentle Castile soap
- Bicarbonate of soda
- Natural clays
- Apple cider vinegar
- Vegetable glycerin
- Fruits and vegetables

Conventional commercial products: These hair-care products may be formulated with the following questionable ingredients, which may cause allergies, sensitivities, damaged hair and dry skin, and – at their worst – are also possibly carcinogenic or otherwise toxic! Some of these may also pose threats to our environment! The following is just a partial list of questionable ingredients:

- Parabens
- Sulphates
- Silicones
- Synthetic dyes
- Synthetic fragrances
- Phthalates
- GMOs (genetically modified organisms)
- Triclosan
- Propylene glycol
- PEG (polyethylene glycol)
- Mineral oil

56

The very best way to wash your hair

Generally, most people will discover that a gentle hair washing a
few times a week is sufficient to keep their hair and scalp clean.
1. Wet your hair with warm water.
2. Apply a small amount of hair cleanser to the palms of
your hands and rub together.
3. Starting at the crown of your head, begin
massaging the cleanser into your scalp and work
your way down to the ends of your hair. Massage for
at least 2 minutes to allow the botanical ingredients
to nourish your hair.
4. Rinse the cleanser out with warm water,
while massaging your scalp to increase
blood circulation.
5. Continue to massage the scalp for 30 more
seconds under warm water.
6. Gently pat your hair dry with a towel.

57

How to condition your hair perfectly

Lightly conditioning your hair after a gentle wash will
moisturise your scalp and help keep your hair smooth,
silky and manageable until your next hair wash.
1. While your freshly cleansed hair is still damp, apply a
generous amount of conditioner to the palms of your hands
and rub together.
2. Starting at the ends of your hair, massage the conditioner into
your hair and scalp.
3. Leave on for 4–5 minutes.
4. Rinse the conditioner out with warm water.
5. (optional step) Apply a small amount of conditioner to the ends of your
hair only and leave on for 30 more seconds before rinsing again.

58

How to deep condition your hair

Deeply conditioning your hair and scalp a few times a month will intensely
nourish your hair and smooth your scalp.
1. While your freshly cleansed hair is still damp, apply a generous amount of
conditioner to the palms of your hands and rub together.
2. Starting at the ends of your hair, massage the deep conditioner into your
hair and scalp.
3. Gently comb the conditioner through your hair from the roots to the tips.
4. Put a shower cap on your head.
5. Leave the conditioner in your hair for at least 30 minutes or, ideally, overnight.
6. Rinse out the deep conditioner with warm water.

59

Tips and tricks for growing long, healthy hair

On average, your hair can grow up to 15 centimetres every year. The secret to healthy, good-looking hair is a combination of nourishing your body on the inside, while nurturing your hair on the outside. Follow the following tried-and-tested tips for beautiful hair.

1. Eat a healthy diet full of foods that contain omega-3 fatty acids, vitamin E, biotin, silica, zinc, antioxidants, protein, sulphur, iron, selenium and folate. Consider taking a vitamin D supplement, too.
2. Cleanse your hair gently and only when necessary.
3. Condition and nourish your hair with plant-based oils and extracts at least once a week to protect the hair follicles.
4. Avoid wearing tight ponytails and hair styles in which the hair shaft is tugged or treated roughly.

5. Trim the ends of your hair on a regular basis. Even snipping off a few millimetres every few months will get rid of any split ends, which, if left alone, will continue to split even further up the hair shaft.
6. Allow your hair to air dry instead of applying harsh heat.
7. Avoid an excessive use of hot styling tools such as curlers, straighteners and crimpers.
8. Remember, it is normal for the average healthy person to shed up to 150 hairs per day.

60

Choosing the best hair-treatment ingredients for your hair type

Every hair type can be perfectly nurtured with specific ingredients. Use this chart to determine the most appropriate ingredients for your individual hair type.

Hair Type	Oils	Essential Oils
Normal	Apricot, argan, baobab, borage seed, grapeseed, jojoba, kukui nut, sweet almond	Cedarwood, clary sage, geranium, lavender, rosemary, sandalwood
Oily	Apricot, argan, hazelnut, jojoba, neem, sesame, shea nut, sunflower	Cypress, geranium, grapefruit, lavender, lemon, orange, peppermint, rosemary, tea tree
Dry	Almond, argan, avocado, baobab, castor, coconut, evening primrose, macadamia nut, meadowfoam seed, neem, olive, rosehip seed, soya bean, tamanu, wheat germ oil	Carrot seed, clary sage, *Eucalyptus citriodora*, Helichrysum, lavender, palmarosa, patchouli, rosemary, sandalwood
Processed & damaged	Argan, avocado, baobab, borage, camelina, castor, evening primrose, hemp seed, macadamia nut, olive, pomegranate, rosehip, sesame, soya bean, vitamin E	Calendula, carrot seed, clary sage, cypress, German chamomile, Helichrysum lavender, myrrh
Coarse & ethnic	Argan, avocado, castor, coconut, hemp seed, macadamia nut, olive, pomegranate seed, rosehip, shea butter, vitamin E	Carrot seed, German chamomile, Helichrysum, sandalwood, tea tree,

▼ Borage

61

Choosing beneficial herbs to care for your hair

The herbs listed in the chart on the right may produce lustre, increase manageability, condition, promote growth and strengthen the hair shaft. Try adding these herbs in unique combinations, such as ginger and peppermint for an energising and stimulating hair rinse. Try thyme and witch hazel together to purify oily hair and scalps.

62

Don't overbrush

Brushing your hair 100 times per day is actually a really bad idea. The hair is stretched and pulled, which can lead to damage and split ends. Instead, use your fingers to massage your scalp for a few minutes to increase blood circulation and relax your mind.

TRY IT

Want classic curls? Then, reach for foam rollers instead of heated ones. Foam rollers are a more gentle way of achieving beautiful bounce and curls, as no heat is involved.

Herb	Botanical Name
Aloe vera	*Aloe vera*
Basil	*Ocimum basilicum*
Black tea	*Camellia sinensis*
Black walnut	*Juglans nigra*
Burdock	*Arctium lappa*
Calendula	*Calendula officinalis*
Cassia	*Cinnamomum burmannii*
Chamomile	*Matricaria chamomilla*
Cloves	*Syzygium aromaticum*
Comfrey	*Symphytum officinale*
Dandelion	*Taraxacum officinale*
Fenugreek	*Trigonella foenum-graecum*
Flaxseed	*Linum usitatissimum*
Ginger	*Zingiber officinale*
Henna	*Lawsonia inermis*
Hibiscus	*Hibiscus sabdariffa*
Horsetail	*Equisetum arvense*
Lavender	*Lavandula intermedia, L. pendunculata, L. officinalis, and L. angustifolia*
Lemon balm	*Melissa officinalis*
Lemongrass	*Cymbopogon citratus*
Licorice root	*Glycyrrhiza glabra*
Mullein	*Verbascum densiflorum*
Nettle	*Urtica dioica*
Oat straw	*Avena sativa*
Peppermint	*Mentha x piperita*
Rosemary	*Rosmarinus officinalis*
Sage	*Salvia officinalis*
Thyme	*Thymus vulgaris*
Watercress	*Nasturtium officinale*
Witch hazel	*Hamamelis virginiana*
Yarrow	*Achillea millefolium*

Hair Cleansers

Hair cleansers are an important step in maintaining clean, shiny hair. Cleansers work to remove dirt, oil and product build-up. Gently cleansing the hair and scalp can turn limp, lifeless hair into squeaky-clean, shiny hair that is ready for styling. Contrary to popular belief, you do not need to wash your hair every day. Unless you have baby-fine hair and/or a very oily scalp, cleansing a few times a week is plenty. Cleansing your hair too frequently can result in split ends, a dry scalp and unruly, parched hair.

63

Choosing and using different types of shampoo

Liquid shampoos: These contain surfactants to remove oils, dirt and odours from the scalp and hair. Natural shampoos do not usually lather and sud as much as conventional cleansers. Nevertheless, they cleanse incredibly well. Liquid shampoos are applied to wet hair and massaged through from the roots to the ends of the hair, and then rinsed with warm water. They can be used a few times a week.

Dry shampoos: These make brilliant 'quick fixes' for dirty hair. Dry shampoos are waterless and contain dry ingredients such as cornflour, arrowroot powder, clay and even rice powder. A small amount is sprinkled over the scalp and either massaged into or combed through the hair to soak up oil, remove dirt and even boost volume. Dry shampoos are regularly perfumed with aromatic essential oils to eliminate odours. They can be used daily.

Bicarbonate of soda: The 'no-shampoo' protocol has become pretty popular recently. Bicarbonate of soda is used to cleanse and purify the hair, and is then followed with a vinegar-based herbal rinse to balance out the pH level (see pages 68–71). Bicarbonate of soda mixed with water (usually in a 1:3 ratio) is gently massaged into the hair and onto the scalp, and then rinsed away with warm water. Since bicarbonate of soda is very alkaline, this particular cleansing method should only be performed once or twice a week.

FIX IT

• When you first start the 'no-shampoo' protocol and choose to cleanse your hair with bicarbonate of soda and condition it with a vinegar-based rinse, you may notice that your hair becomes either super oily or super dry. Stick with it until your hair adjusts and finds a happy balance. It can take up to one month before you start to enjoy natural results.

• Castile-soap-based cleansers can sometimes leave the hair feeling parched and hard to comb through. Try working the cleanser though your hair with flat hands in a pressing motion instead of a massaging one with your fingers. In this way, the hair will stay flat instead of becoming tangled while it is wet.

 64

Plan your packaging

Don't get caught in the trap of making a gorgeous natural hair cleanser with no thought to how you will store it. Recycle used shampoo and conditioner bottles to hold your handcrafted creations. Simply wash and rinse them out well and shake a tablespoon of alcohol through them to sanitise.

 65

Make a basic liquid shampoo

Liquid Castile soap has been around for a very long time. It is made by saponifying oil with potassium hydroxide, which is an alkaline ingredient. Diluted liquid Castile soap is wonderful for naturally cleansing the hair. Many of the liquid Castile soaps that are available also contain natural essential oils such as orange, lemon, lime, peppermint and even eucalyptus. Adding essentials oils in this way gives the shampoo a fresh smell and sometimes provides an invigorating cleansing experience, as in the case of a peppermint Castile soap.

Ingredients:
• *240 ml water*
• *80 ml liquid Castile soap*
Makes approx. 320 ml

Combine both of the ingredients in a pourable bottle, and mix well.

To use: Apply a small amount to wet hair to cleanse. Rinse with warm water.

 66

Make a basic dry shampoo

Powdered, dry natural ingredients, such as rice powder, arrowroot powder, cornflour, oat powder, tapioca starch, kaolin clay, cocoa powder, horsetail powder, orris root powder and bicarbonate of soda can be combined with small amounts of essential oil to cleanse the hair without using any water in order to make the hair appear thicker and fuller.

Ingredients:
• *30 g arrowroot powder*
• *1 tablespoon bicarbonate of soda*
• *20 drops essential oil*
Makes approx. 40 g

Place all of the ingredients into a blender or food-processor, and process until well combined. Transfer to a sugar shaker with a lid.

To use: Sprinkle a teaspoon-sized amount into your hair. Use your fingers to massage your hair from the scalp to the roots. Use a bristle brush to comb through your hair evenly. Style as desired.

Recipes:
Six of the Best Hair Cleansers

67

Splendid Lavender & Grapefruit Voluminous Cleansing Shampoo for Oily Hair

Makes approx. 250 ml
Best for normal and oily hair

Start your day with the aromatic aroma of herbaceous lavender and awakening grapefruit, while you softly cleanse your hair to create body and bounce.

120 ml liquid Castile soap
60 ml witch hazel extract
60 ml lavender floral water
1 tablespoon fine sea salt
2 tablespoons rosemary tincture
60 drops lavender essential oil
30 drops grapefruit essential oil

1. Measure out all of the ingredients into a pourable bottle or a recycled shampoo bottle and then shake well to dissolve the salt. Store in the fridge and use within 2 weeks.

To use: Shake the bottle well. Apply a small amount to wet hair and gently massage through the hair from the roots to the ends. Rinse thoroughly with warm water and repeat if desired. This shampoo does not lather and sud as much as a conventional high-sudsing shampoo. Follow with a conditioner or conditioning hair rinse.

68

Burdock & Ginger Infused Cleansing Shampoo for a Flaky Scalp

Makes approx. 250 ml
Best for an itchy, flaky scalp and dry hair

Cleanse and invigorate your hair and scalp with this peppermint- and tea-tree-infused natural shampoo, which is made with apple cider vinegar, burdock seed and freshly grated ginger.

120 ml raw apple cider vinegar
1 tablespoon ground burdock seed
2 teaspoons freshly grated ginger
120 ml liquid Castile soap
2 tablespoons argan oil
30 drops tea tree essential oil
10 drops palmarosa essential oil
10 drops clary sage essential oil
5 drops peppermint essential oil

1. Add the raw apple cider vinegar, burdock seed and grated ginger to a small saucepan with a tight-fitting lid.
2. Simmer on a very low heat for 1 hour.
3. Remove from the heat and keep the lid on the saucepan while the liquid cools to room temperature for about 1 hour.
4. Strain the liquid from the spent herbs into a pourable bottle or recycled shampoo bottle.
5. Add the liquid Castile soap, argan oil and essential oils. Shake well.

To use: Shake the bottle well. Apply a small amount to wet hair and gently massage through the hair from the roots to the ends. Leave on for 2 minutes. Rinse thoroughly with warm water and repeat if desired. This shampoo does not lather and sud as much as a conventional high-sudsing shampoo. Follow with a conditioner or conditioning hair rinse.

69

Exquisite Bulgarian Rose Hair Powder

Makes approx. 280 g
Best for all hair types

A refreshing dry cleanser to use between wet washes, to simply add volume and body, or to quickly polish dry hair. Soak up scalp oil and impart a delightful rosy fragrance to your tresses with a sprinkle of this wonderful and refreshing hair powder.

60 g arrowroot powder
2 tablespoons kaolin clay
2 tablespoons bicarbonate of soda
1 tablespoon rice powder
12 drops Bulgarian rose essential oil

1. Place all of the ingredients in a blender or food-processor and process until well combined. Transfer to a sugar shaker with a lid.

To use: Sprinkle a teaspoon-sized amount into your hair. Use your fingers to massage the powder through your hair from the scalp to the roots. Use a bristle brush to comb through your hair evenly. Style as desired.

Sweet-Scented Cocoa & Vanilla Dark Hair Powder

Makes approx. 75 g
Best for darker-coloured hair

This is a refreshing and delightful-smelling hair powder. The cocoa powder enhances the natural colour and beauty of darker hair types.

25 g cocoa powder
30 g cornflour
2 tablespoons bicarbonate of soda
40 drops vanilla absolute

1. Place all of the ingredients into a blender or food-processor and process until well combined. Transfer to a sugar shaker with a lid.

To use: Sprinkle a teaspoon-sized amount into your hair. Use your fingers to massage through your hair from the scalp to the roots. Use a bristle brush to comb through your hair evenly. Style as desired.

Herbal Rosemary & Mint Shampoo

Makes approx. 350 ml
Best for all hair types

This remarkable cleansing shampoo will leave your hair healthy, shiny and gorgeous. The jojoba oil and meadowfoam seed oil work to fortify the hair follicles. Rosemary and spearmint essential oils support a healthy scalp and also smell marvelous!

120 ml liquid Castile soap
60 ml Aloe vera gel
60 ml chamomile floral water
2 tablespoons vegetable glycerin
1 tablespoon comfrey tincture
1 tablespoon rosemary tincture
1 tablespoon calendula tincture
1 tablespoon finely grated sea salt
1 tablespoon jojoba oil
1 tablespoon meadowfoam seed oil
1 teaspoon vitamin E oil
40 drops rosemary essential oil
20 drops spearmint essential oil

1. Measure out all of the ingredients into a pourable bottle or a recycled shampoo bottle, and shake well to dissolve the salt. Store in the fridge and use within 2 weeks.

To use: Shake the bottle well. Apply a small amount to wet hair and gently massage through the hair from the roots to the ends. Rinse thoroughly with warm water and repeat if desired. This shampoo does not lather and sud as much as a conventional high-sudsing shampoo. Follow with a conditioner or conditioning hair rinse.

Clarifying Clary Sage, Oat & Bicarbonate of Soda Cleanser

Makes approx. 115 g
Best for all hair types

In this cleanser, the bicarbonate of soda works to purify the hair and exfoliate the scalp, the soothing oat flour works to soothe and protect, while the clary sage essential oil helps to normalise the scalp.

4 tablespoons bicarbonate of soda
2 tablespoons finely ground oat flour
40 drops clary sage essential oil

1. Place all of the ingredients into a small food-processor or blender and process until well blended. Transfer to a shaker container with a lid.

To use: Sprinkle a generous amount over wet hair and massage into your scalp and hair. Leave on for 2 minutes and rinse out with warm water. Condition as usual.

Hair Conditioners

Hair conditioners contain hydrating plant oils that work to restore the softness and lustre that shampoos can cleanse away. They help create soft, manageable hair that is easy to comb and style. Being high in essential fatty acids, they can tame even the most unruly, frizzy locks into a silky, intensely conditioned and beautified head of healthy-looking hair.

TRY IT

If you have baby-fine or oily hair, simply focusing on conditioning the ends of your hair, instead of your scalp, may keep your hair from becoming too oily.

73

Choosing and using different types of conditioner

Regular conditioners: These can protect hair colour, quench dryness, smooth the cuticles, de-frizz and detangle. They are applied after the hair has been cleansed. You simply massage the conditioner in from the roots to the tips. Conditioners are left on for one to three minutes and rinsed out with warm water.

Deep conditioners and hair packs: These treatments work to pamper your stressed-out locks, seal split ends, enhance curls, nourish the scalp and thicken baby-fine hair. They have a high concentration of nutritious oils that coat and penetrate the hair shaft to smooth out and balance. They are left on for up to one hour and then rinsed out with warm water. Deep conditioners should be used weekly.

Leave-in conditioners: These can be misted through the hair or massaged in after cleansing. Leave-in conditioners improve the texture of the hair and allow for hassle-free combing and styling.

Hair oils and serums: These are very concentrated potions of pure plant oils and extracts. They are waterless formulas, which are made with beneficial plant oils and essential oils that hydrate, de-frizz, seal split ends and impart a stunning shine to the hair. Often, the oils are infused with hair-helping herbs such as burdock root and rosemary. Hair oils are very concentrated and a little bit goes a very long way!

74

Less is more

Hair oils and serums are very concentrated and can weigh down the hair if too much is applied. Apply a tiny drop to the palms of your hands and rub together. Gently massage into the ends of your hair and then work your way up to your scalp. Brush though with a soft bristle brush to distribute the oils evenly through your hair.

75

Unusual remedies

• Mayonnaise really does work to condition the hair. Work a tablespoon or two through damp hair, leave on for 5 minutes and shampoo as usual.

• If you are out and about and notice pesky flyaway hairs or bothersome static, reach for your hand cream. Apply a small amount to the unruly hairs to keep them down.

76

Make a basic natural hair conditioner

Ingredients:
- *2 tablespoons jojoba oil*
- *2 teaspoons emulsifying wax NF*
- *½ teaspoon stearic acid*
- *½ teaspoon liquid lecithin*
- *120 ml distilled water*
- *30 drops lavender essential oil*

Makes approx. 160 ml

1. In a heat-safe glass measuring jug, combine the jojoba oil, emulsifying wax, stearic acid and liquid lecithin. Place the measuring jug in a saucepan containing a few centimetres of simmering water.

2. In another heat-safe glass measuring jug, add in the distilled water. Place this jug in another saucepan containing a few centimetres of simmering water.

3. Heat both mixtures to a temperature of 71°C (160°F) until the oil and wax have melted together. Carefully remove both measuring jugs from the simmering water.

4. Pour the oil/wax mixture into a mixing bowl and begin mixing with a hand-held mixer set to low. Blend the water mixture into the oil/wax on low for 1 minute. Bring the speed up to high and continue mixing for 5 minutes or until the temperature is below 38°C (100°F).

5. Add the lavender essential oil and mix again on high for a further 5 minutes. Transfer the mixture to a sanitised, pourable bottle or a recycled, clean conditioner bottle. Store in the fridge and use within 2 weeks.

To use: After shampooing your hair, apply generously from the roots to the ends and leave on for 2–5 minutes. Rinse well with warm water.

Recipes:
Twelve of the Best Hair Conditioners

77

Honey & Sage Hair Conditioner

Makes approx. 175 ml
Best for all hair types

Fortify your hair with this honey and sage hair conditioner. Sweet almond oil protects and nourishes the hair follicles, while the sweet sage provides a refreshing scent.

2 tablespoons sweet almond oil
2 teaspoons emulsifying wax NF
½ teaspoon stearic acid
½ teaspoon liquid lecithin
120 ml distilled water
1 tablespoon honey
⅛ teaspoon xanthan gum powder
30 drops sage essential oil

1. In a heat-safe glass measuring jug, combine the sweet almond oil, emulsifying wax, stearic acid and liquid lecithin.
2. Place the measuring jug in a saucepan containing a few centimetres of simmering water.
3. In another heat-safe glass measuring jug, combine the distilled water, honey and xanthan gum powder. Place this measuring jug in another saucepan containing a few centimetres of simmering water.
4. Heat both mixtures to a temperature of 71°C (160°F) until the oil and wax have melted together. Carefully remove both measuring jugs from the simmering water.
5. Pour the oil/wax mixture into a mixing bowl and begin mixing with a hand-held mixer set to low. Blend the water mixture into the oil/wax mixture on low for 1 minute.
6. Bring the speed up to high and continue mixing for 5 minutes or until the temperature is below 38°C (100°F).
7. Add the sage essential oil and mix again on high for a further 5 minutes.
8. Transfer the mixture to a sanitised, pourable bottle or a recycled, clean conditioner bottle. Store in the fridge and use within 2 weeks.

To use: After shampooing your hair, apply the conditioner generously from the roots to the ends and leave on for 2–5 minutes. Rinse well with warm water.

78

Softening Sunflower & Sweet Marjoram Hair Mask

Makes approx. 200 ml
Best for dry, damaged hair

This softening and nourishing hair mask helps rejuvenate your hair with shine. Give your thirsty hair some much-needed moisture with the soothing benefits of natural plant oils and enjoy the relaxing scent of sweet marjoram.

60 ml sunflower oil
3 teaspoons emulsifying wax NF
½ teaspoon stearic acid
½ teaspoon liquid lecithin
120 ml distilled water
⅛ teaspoon xanthan gum powder
30 drops sweet marjoram essential oil
1 teaspoon vitamin E oil

1. In a heat-safe glass measuring jug, combine the sunflower oil, emulsifying wax, stearic acid and liquid lecithin.
2. Place the measuring jug in a saucepan containing a few centimetres of simmering water.
3. In another heat-safe glass measuring jug, combine the distilled water and xanthan gum powder. Place this measuring jug in another saucepan containing a few centimetres of simmering water.
4. Heat both mixtures to a temperature of 71°C (160°F) until the oil and wax have melted together. Carefully remove both measuring jugs from the simmering water.
5. Pour the oil/wax mixture into a mixing bowl and begin mixing with a hand-held mixer set to low. Blend the water mixture into the oil/wax mixture on low for 1 minute.
6. Bring the speed up to high and continue mixing for 5 minutes or until the temperature is below 38°C (100°F).
7. Add the sweet marjoram essential oil and vitamin E oil, and mix again on high for a further 5 minutes.
8. Transfer the mixture to a sanitised, pourable bottle or a recycled, clean conditioner bottle. Store in the fridge and use within 2 weeks.

To use: After shampooing your hair, apply the hair mask generously from the roots to the ends and leave on for 20–30 minutes. Rinse well with warm water.

79

Peppermint & Tea Tree Leave-In Conditioner

Makes approx. 300 ml
Best for all hair types

This invigorating, scalp-tingling leave-in conditioner helps to smooth even the most unruly hair into shiny and gorgeous tresses. A little goes a long way!

2 tablespoons coconut oil
2 tablespoons jojoba oil
3 teaspoons emulsifying wax NF
½ teaspoon stearic acid
½ teaspoon liquid lecithin
240 ml peppermint floral water
⅛ teaspoon xanthan gum powder
20 drops tea tree essential oil
5 drops peppermint essential oil

1. In a heat-safe glass measuring jug, combine the coconut oil, jojoba oil, emulsifying wax, stearic acid and liquid lecithin.
2. Place the measuring jug in a saucepan containing a few centimetres of simmering water.
3. In another heat-safe glass measuring jug, combine the peppermint floral water and xanthan gum powder. Place this measuring jug in another saucepan containing a few centimetres of simmering water.
4. Heat both mixtures to a temperature of 71°C (160°F) until the oils and wax have melted together. Carefully remove both measuring jugs from the simmering water.
5. Pour the oil/wax mixture into a mixing bowl and begin mixing with a hand-held mixer set to low. Blend the water mixture into the oil/wax mixture on low for 1 minute.
6. Bring the speed up to high and continue mixing for 5 minutes or until the temperature is below 38°C (100°F).
7. Add the tea tree and peppermint essential oils, and mix again on high for a further 5 minutes.
8. Transfer the mixture to a sanitised, pourable bottle or a recycled, clean conditioner bottle. Store in the fridge and use within 2 weeks.

To use: Shake the container well. After shampooing your hair, apply a small amount of conditioner to towel-dried hair and massage in. Style as usual.

80

Moisturising Treacle Mask

Makes 1 application
Best for dry hair

This moisturising hair mask will nourish and soothe every single strand of suffering hair.

175 g black treacle
1 egg yolk
50 g mashed avocado

1. Mix all of the ingredients into a thick paste.

To use: Massage into damp hair from the roots to the tips. Cover with a shower cap and leave on for 30 minutes to 1 hour. Rinse out with warm water and shampoo if desired.

81

Boozy & Bouncy Hair Conditioning Rinse

Makes 1 application
Best for fine hair

This simple conditioning rinse of tepid beer and rum mixed with a little bit of lime juice will leave your hair shiny and bouncy.

240 ml warm beer
2 tablespoons rum
1 tablespoon lime juice

1. Mix all of the ingredients together well in a glass or pourable jug.

To use: After cleansing your hair, slowly pour the rinse though your hair and massage from the roots to the tips. Leave on for 2 minutes and rinse with warm water.

82

Hungry Hair Nourishing Oil

Makes approx. 60 ml
Best for dry, damaged, processed and coarse hair

This wonderful and nourishing hair oil is loaded with vitamins A, D and E and will keep your hair youthful and healthy-looking. There is no water in this formula and a little goes a long way in nurturing your scalp and hair.

1 tablespoon avocado oil
1 tablespoon apricot
  kernel oil
1 tablespoon wheat germ oil
1 tablespoon olive oil
10 drops ylang ylang
  essential oil
10 drops petitgrain
  essential oil
5 drops neroli essential oil

1. Combine all of the ingredients in a dropper bottle. Shake the bottle well.

To use: Massage a few teaspoons into your hair. Comb through and leave on for 15 minutes. Shampoo and condition as usual.

83

Ocean Mist for Texture and Volume

Makes approx. 200 ml
Best for curly or wavy hair

Give your lovely locks the look of frolicking at the beach all day. This leave-in spray creates texture and volume to enhance natural waves and curls. The intoxicating aromas of jasmine and rose softly scent your hair.

180 ml rose floral water
2 tablespoons Epsom salts
1 tablespoon vodka
6 drops jasmine absolute

1. Place the floral water and Epsom salts in a small saucepan.
2. Warm over a very low heat until the salt has dissolved.
3. Cool to room temperature.
4. Pour the vodka into a spray bottle and add the jasmine absolute. Shake the bottle to combine.
5. Add the salt mixture to the bottle and shake well. Store in the fridge and use within 2 weeks.

To use: Spray the mist onto clean, towel-dried hair and allow the hair to dry naturally. If you use the mist frequently, make sure that you treat your hair to a deep-conditioning treatment to avoid it drying out as a result of the salt.

84

Healing & De-Frizzing Dry Hair Serum

Makes approx. 30 ml
Best for dry, damaged and frizzy hair

This split-end-mending hair serum will control bothersome frizzes and help promote soft, silky hair after just one use.

3 teaspoons jojoba oil
1 teaspoon olive oil
1 teaspoon evening primrose oil
½ teaspoon flax seed oil
¼ teaspoon castor oil
2 drops lavender essential oil
2 drops cedarwood essential oil
2 drops rosemary essential oil

1. Combine all of the ingredients into a dropper bottle. Shake the bottle well.

To use: Warm up a few drops of the serum in the palms of your hands and massage through damp or dry hair. Apply one more drop to the ends of your hair if needed. Style as usual.

85

Harmonising Neem & Calendula Hair Tonic

Makes approx. 150 ml
Best for normal, dry and damaged hair

This delightful, leave-in tonic works to balance the scalp and encourage healthy hair growth. Neem leaf is commonly used in hair-care products throughout India.

60 ml distilled water
1 tablespoon neem leaf powder
60 ml calendula tincture
2 tablespoons rosemary tincture
10 drops myrrh essential oil
5 drops ylang ylang essential oil

1. Bring the distilled water to a boil and add the neem leaf powder. Cover the pan with a tight-fitting lid.
2. Allow the liquid to cool to room temperature and then strain the liquid from the herb into an amber-coloured glass bottle with a lid.
3. Add the calendula tincture, rosemary tincture and essential oils. Shake well. Store in the fridge and use within 2 weeks.

To use: Apply drops of the hair tonic directly to your scalp and massage gently or comb through. Use 3–4 times per week.

86

Neroli & Citrus Hair Balm

Makes approx. 60 ml
Best for split ends and flyaway hair

Got frizz? Got split ends? Handcraft this moisturising and smoothing hair balm that is scented with the intoxicating aromas of neroli and citrus. Just a small amount goes a long way to controlling troublesome hair!

2 teaspoons grated beeswax
1 tablespoon coconut oil
2 teaspoons shea butter
2 teaspoons sweet almond oil
2 teaspoons wheat germ oil
¼ teaspoon vitamin E oil
30 drops sweet orange essential oil
10 drops neroli essential oil

1. Place the beeswax, coconut oil, shea butter, sweet almond oil and wheat germ oil in a small glass measuring jug. Place the measuring jug in a saucepan of simmering water.
2. Allow the ingredients to melt over a low heat. Remove from the heat and stir in the vitamin E oil and essential oils.
3. Transfer the mixture to a heat-safe container with a lid.

To soothe and moisturise: Massage a small amount of the balm in the palms of your hands and gently apply to dry hair to smooth and shine.

To deep condition: Massage a few teaspoons into your hair. Comb through and leave on for 15 minutes. Shampoo and condition as usual.

▲ Lustre & Shine Double Avocado Moisturising Hair Mask and Coconut Milk & Rosemary Hair Mask

87

Lustre & Shine Double Avocado Moisturising Hair Mask

Makes 1 treatment
Best for dry and damaged hair

This is a very intense hair hydrator to help rejuvenate dry, damaged hair. Your lovely locks will be left shiny, soft and luxurious!

1 tablespoon avocado oil
50 g mashed avocado
2 egg yolks
20 drops lemon essential oil

1. Mix all of the ingredients in a small bowl.

To use: Apply to damp hair and massage from the roots to the ends. Place a shower cap on your head and allow the mask to work for 1 hour. Shampoo as usual.

88

Coconut Milk & Rosemary Hair Mask

Makes approx. 180 ml
Best for all hair types

This amazing hair mask is fortified with nourishing coconut milk and rosemary. After just one use, your hair is left feeling silky and smooth.

180 ml canned unsweetened coconut milk
3 tablespoons freshly chopped rosemary leaves
1 teaspoon vitamin E oil

1. Place the coconut milk and rosemary leaves in a small saucepan with a lid. Simmer over a low heat for 25 minutes, stirring frequently so as not to burn the coconut milk.
2. Strain the infused coconut milk from the herb and into a small glass container.
3. Stir in the vitamin E oil.
4. Allow to cool to room temperature. Store in the fridge and use within 10 days.

To use: Cleanse your hair as normal and towel dry. Massage in half of the hair mask and comb through your hair. Place a shower cap on your hair and allow the mask to work its magic for at least 1 hour or, ideally, overnight. Rinse your hair with warm water for several minutes and style as usual.

Herbal Hair Rinses

Herbal rinses and clarifiers are restorative, usually crafted with herbal infusions and extracts, and often contain citrus juice or apple cider vinegar to balance the hair's natural pH. They do not contain surfactants like liquid shampoos and do not lather. They are terrific at freeing the hair of dulling build-up, excess oils and odours. Some herbal hair rinses are designed to be colour enhancers to bring out your hair's natural colour and highlights. They can be used a few times a week. Herbal rinses are popular among those who adhere to the 'no-shampoo' protocol.

The best way to apply a herbal rinse is via a spray bottle. This way, you can thoroughly coat the hair and scalp, and then massage the rinse through. Herbal-infused hair rinses can normally be left in, but ones made with citrus juice or vinegar need to be rinsed out with warm water.

89

Make a basic vinegar hair rinse

Ingredients:
• *80 ml water*
• *1 tablespoon apple cider vinegar*
Makes 1 application

1. Combine both of the ingredients in a spray bottle.

To use After cleansing your hair, spray the hair rinse through your hair and massage from the roots to the ends. Leave on for several minutes and then rinse out and style as usual.

90

Make a basic herbal-infused hair rinse

Ingredients:
• *120 ml boiling water*
• *1 tablespoon dried herb (see the charts on the opposite page for suggestions on which herbs to use)*
Makes 1 application

1. Place the herb in a saucepan of boiling water. Turn off the heat and allow the herb to infuse into the water.
2. Cool to room temperature before transferring to a spray bottle. Please note that roots such as burdock and comfrey will need to steep for longer – overnight is best.

To use: After cleansing your hair, spray the hair rinse through your hair and massage from the roots to the ends. Leave on for several minutes and either rinse out or leave in and style as usual.

TRY IT

• Add a few drops of essential oil to your hair rinses. Tea tree is great for a flaky scalp, Helichrysum is good for dry hair, and black pepper is good for oily hair.

• If you have dry hair or a dry scalp, try adding a teaspoon of oil to your hair rinse and massaging this into your scalp and hair before rinsing.

91

Choosing herbs for colour-enhancing rinses

For very subtle and nourishing colour-enhancing effects, select a combination of these recommended herbs for your particular hair colour.

Hair Colour	Recommended Herbs
Blonde	Calendula, chamomile, grapefruit peel, lemon peel, mullein, oat straw, rhubarb root
Brunette	Black tea, black walnut hulls, cloves, comfrey root, nettle, rosemary, sage
Red	Alkanet root, calendula, cinnamon, hibiscus, red clover, rosehips, turmeric

92

Choosing herbs for problem hair

Treat and support problem hair and scalp with a combination of these restorative herbs that may promote a healthy scalp and manageable hair.

Hair Problem	Recommended Herbs
Flaky scalp	Burdock root, chamomile, green tea, lavender, mugwort, nettle, oregano, peppermint, rosemary, spearmint, thyme
Thinning hair	Basil, comfrey, hops, horsetail, lavender, nettle, peppermint, rosemary
Oily hair & scalp	Calendula, horsetail, lavender, lemon balm, lemon peel, rosemary, witch hazel

93

Boost your rinses

Replace the water in the recipe for a herbal hair rinse with an equal amount of floral water (hydrosol) for added benefits.

Recipes:
Eight of the Best Herbal Hair Rinses

▶ Dried hibiscus flowers

94

Lovely Lime & Chamomile Hair Rinse

Makes 1 application
Best for all hair types

Give your hair a shiny boost with this fresh lime juice and chamomile hair rinse.

1 tablespoon dried chamomile flowers
1 tablespoon dried horsetail
1 tablespoon dried oat straw
240 ml boiling water
Juice of 1 fresh lime

1. Place the herbs in a small saucepan and add the boiling water.
2. Cover the saucepan with a lid and allow the herbs to infuse the water. Cool to room temperature.
3. Strain the liquid from the spent herbs and place in a spray bottle along with the fresh lime juice.

To use: After cleansing, spray all of the hair rinse into your hair and then massage through your hair and into your scalp. Leave on for 5 minutes and then rinse with warm water or leave on and style as usual.

95

Earl Grey & Espresso Dark Hair Rinse

Makes 1 application
Best for dark hair

This delightful, colour-enhancing hair rinse will bring out the natural highlights of dark hair.

240 ml boiling water
2 teaspoons black walnut hulls
1 teaspoon ground cloves
1 Earl Grey tea bag
2 teaspoons instant espresso powder

1. Steep the herbs, tea bag and espresso powder in the boiling water until cooled to room temperature.
2. Remove the tea bag and dispense the hair rinse in a spray bottle.

To use: After cleansing, spray all of the hair rinse into your hair and then massage through your hair and into your scalp. Leave on for 5 minutes and then rinse with warm water and style as usual.

96

Calendula & Chamomile Blonde Hair Rinse

Makes 1 application
Best for blonde hair

Blondes will have a lot of fun with this colour-enhancing hair rinse, which brings out natural golden highlights.

240 ml boiling water
2 tablespoons dried chamomile flowers
2 tablespoons dried calendula flowers
1 tablespoon freshly grated lemon zest

1. Steep the herbs and lemon zest in the boiling water until cooled to room temperature.
2. Strain the liquid from the spent herbs and lemon zest.
3. Dispense the hair rinse into a spray bottle.

To use: After cleansing, spray all of the hair rinse into your hair and then massage through your hair and into your scalp. Leave on for 5 minutes and then rinse with warm water or leave on and style as usual.

97

Happy Hibiscus & Cinnamon Red Hair Rinse

Makes 1 application
Best for red hair

Your natural red hair will be happy with this wonderful, colour-enhancing hair rinse, which brings out the natural loveliness of red hair.

240 ml boiling water
2 tablespoons dried hibiscus flowers
1 tablespoon dried calendula flowers
2 teaspoons dried rosehips
1 teaspoon ground cinnamon bark

1. Steep the herbs and cinnamon bark in the boiling water until cooled to room temperature.
2. Strain the liquid from the spent herbs.
3. Dispense the hair rinse into a spray bottle.

To use: After cleansing, spray all of the hair rinse into your hair and then massage through your hair and into your scalp. Leave on for 5 minutes and then rinse with warm water or leave on and style as usual.

◀ Dried chamomile flowers

98

Herbal Rinse for Normal Hair

Makes 1 application
Best for normal hair

Keep your hair in tip-top shape with the purifying benefits of apple cider vinegar and hair-friendly herbs.

480 ml boiling water
3 tablespoons apple cider vinegar
1 teaspoon dried parsley
1 teaspoon dried rosemary
1 teaspoon dried horsetail
1 teaspoon dried oat straw

1. Bring the water to a boil in a small saucepan. Add the vinegar and herbs, cover with a lid and turn off the heat.
2. Allow the herbs to infuse in the water and vinegar until cooled to room temperature.
3. Strain the liquid from the spent herbs.
4. Dispense the rinse into a spray bottle for the best results.

To use: After cleansing your hair, spray or pour the hair rinse through your hair and massage in from the roots to the tips. Rinse with warm water or, for the best results, leave in and style as usual. Do not drink!

◀ Dried oat straw

◀ Dried burdock root

99

Herbal Rinse for Oily Hair

Makes 1 application
Best for oily hair

The combination of astringent witch hazel and apple cider vinegar will purify your hair and scalp and keep your oil production in check.

480 ml boiling water
3 tablespoons apple cider vinegar
1 teaspoon dried lemon balm
1 teaspoon dried witch hazel
1 teaspoon dried lavender
1 teaspoon dried chopped grapefruit peel
1 teaspoon dried rosemary

1. Bring the water to a boil in a small saucepan. Add the vinegar and herbs, cover with a lid and turn off the heat.
2. Allow the herbs to infuse in the water and vinegar until cooled to room temperature.
3. Strain the liquid from the spent herbs.
4. Dispense the rinse into a spray bottle for the best results.

To use: After cleansing your hair, spray or pour the hair rinse through your hair and massage in from the roots to the tips. Rinse with warm water or, for the best results, leave in and style as usual. Do not drink!

100

Herbal Rinse for Dry & Coarse Hair

Makes 1 application
Best for dry or coarse hair

The calming herbs in this formula will provide comfort to a dry scalp and hair.

480 ml of boiling water
2 tablespoons apple cider vinegar
1 teaspoon dried calendula
1 teaspoon dried chamomile
1 teaspoon dried burdock root
1 teaspoon dried sage
1 teaspoon dried mullein
1 teaspoon dried nettle
2 teaspoons vegetable glycerin

1. Bring the water to a boil in a small saucepan. Add the vinegar and herbs, cover with a lid and turn off the heat.
2. Allow the herbs to infuse in the water and vinegar until cooled to room temperature.
3. Strain the liquid from the herbs. Mix in the vegetable glycerin.
4. Dispense the rinse into a spray bottle for the best results.

To use: After cleansing your hair, spray or pour the hair rinse through your hair and massage in from the roots to the tips. Rinse with warm water or, for the best results, leave in and style as usual.

101

Herbal Rinse for Processed & Damaged Hair

Makes 1 application
Best for processed or damaged hair

Gently restore your damaged hair by using this rinse that is infused with nutritious and restoring herbs.

480 ml boiling water
2 tablespoons apple cider vinegar
1 teaspoon dried calendula
1 teaspoon dried marshmallow root
1 teaspoon dried burdock root
1 teaspoon dried mullein
1 teaspoon dried chia seed
2 teaspoons vegetable castor oil

1. Bring the water to a boil in a small saucepan. Add the vinegar and herbs, cover with a lid and turn off the heat.
2. Allow the herbs to infuse in the water and vinegar until cooled to room temperature.
3. Strain the liquid from the herbs. Mix in the vegetable castor oil.
4. Dispense the rinse into a spray bottle for the best results.

To use: After cleansing your hair, spray or pour the hair rinse through your hair and massage in from the roots to the tips. Rinse with warm water or, for the best results, leave in and style as usual. Do not drink!

4 Fabulous Facial Care

Plant-based botanical skincare can keep a perfect complexion in tip-top shape but, then again, if you have other beauty worries, there will be an excellent and effective natural treatment for those as well. With the influence of healing herbs, beneficial oils and powerful plant extracts, you may formulate and make to order a perfect product that can lessen the signs of aging, soothe and put a stop to pesky blemishes, deep clean and exfoliate pores, and calm distressed and dry skin by restoring proper levels of hydration.

Facial Cleansers

Gently cleansing your face with a natural, plant-based cleansing formula can remove makeup, sebum build-up and impurities from the skin. When you harness the power of botanicals, you purify your skin in a soothing and nourishing way, which allows its natural healthy glow and vitality to shine through.

102

Choosing a facial cleanser

Water-based cleansers: These mild, herb-infused, water-based cleansers softly bathe and freshen up the skin, leaving it bright and dirt-free. While water-based cleansers may not take away makeup as well as a soap- or oil-based cleanser, they are perfect for those with a dry or sensitive skin type, where a soothing rinse is all that is desired. Nourishing and mild botanical ingredients, such as floral waters, herbs, tinctures, fruit juices, apple cider vinegar and even honey, can all be added to your custom-made formula.

Soap-based cleansers: Containing a natural surfactant, usually liquid Castile soap, these cleansers free the skin of pore-clogging impurities, makeup and oily sebum accumulation. Soap-based cleansers are effortlessly personalised to suit each skin type's specific needs. You can add particular ingredients such as clay for a deep cleanse; sugar or almond meal for an exfoliating cleanse; detoxifying essential oils can be added for those with blemish-prone skin; Aloe vera can be included for those with dry skin; and plant oils such as jojoba oil can even be added for dry skin types.

Oil-based cleansers: These are formulated with nutritious plant oils and beneficial essential oils. Oil-based cleansers are concentrated and do not contain any water. They are superb for removing stubborn makeup – even eye makeup! They can transform parched, mature and sensitive skin into softly cleansed, supple skin. Even oily skin types can use an oil-based cleanser with perfect results.

> **FIX IT**
>
> Because these formulas do not contain any preservatives, it is best to either store them in the fridge or, at the very least, to make sure that you don't touch the product with your fingers or a dirty utensil. A sanitised, pourable bottle with a tight-fitting cap works best.

103
Cleansing with oil-based products

The best way to use an oil-based facial cleanser is to apply a generous amount to dry skin, massage in and remove with a soft cloth. You may rinse with warm water if desired.

104
Cleansing with water-based products

To use a water-based cleanser, you can either spray a generous amount over your face, or apply the required amount to a cotton pad and softly swipe the solution over your skin to freshen it up.

105
Cleansing with soap-based products

Simply apply a small amount of the soap-based cleanser to wet skin, massage into the skin with circular motions and rinse well. Avoid getting soap-based cleansers in your eyes.

106
Mix it up

If you have sensitive or dry skin, use a soap-based cleanser before bed and then switch to a water- or oil-based cleanser in the morning.

107
Double the power

You may need to cleanse your skin twice with a natural cleanser if you are wearing heavily applied cosmetics. Try cleansing with an oil-based cleanser first in order to break up and remove the makeup, and then follow up with a soap-based cleanser.

TRY IT

For makeup-free skin that only requires a light cleanse, pour some warm floral water onto a flannel to saturate. Firmly, yet gently, massage your skin with the flannel to lightly exfoliate and loosen pore-clogging debris.

Recipes:
Twelve of the Best Facial Cleansers

108

Basic Water-Based Cleanser

Makes approx. 200 ml
Best for normal, dry, sensitive or mature skin

A water-based cleanser will whisk away impurities without stripping away essential moisture.

60 ml chamomile floral water
60 ml Aloe vera gel
60 ml vegetable glycerin
2 tablespoons witch hazel extract

1. Combine all of the ingredients in a sanitised bottle. Shake the bottle well. Store in the fridge and use within 2 weeks.

To use: Apply a generous amount of the cleanser to your skin using a spray bottle or cotton pad, massage in with your fingertips and tissue or rinse off.

109

Basic Soap-Based Cleanser

Makes approx. 75 ml
Best for normal, oily, combination or blemish-prone skin

This cleanser loosens dirt, sebum and makeup to leave your skin feeling clean and refreshed.

3 tablespoons boiling water
2 teaspoons plain table salt
4 tablespoons liquid Castile soap

1. Pour the boiling water into a small glass measuring jug. Add the salt and stir well to dissolve. Set aside.
2. Add the liquid Castile soap to a sanitised bottle with a lid or pump.
3. Place a funnel in the top of the bottle and add 1 tablespoon of the salt solution to the liquid Castile soap. Discard any remaining salt solution.
4. Put the lid on the bottle and shake well to thicken. Store in the fridge and use within 2 weeks.

To use: Shake the bottle well and apply a pea-sized amount to wet skin, massage in (avoiding your eyes) and rinse off with warm water.

110

Basic Oil-Based Cleanser

Makes approx. 250 ml
Best for all skin types

This cleanser removes makeup, dirt and skin impurities, while also deeply moisturising and comforting all skin types.

240 ml extra virgin olive oil
2 tablespoons jojoba oil

1. Combine both the ingredients in a small, dry, sanitised bottle. Shake the bottle well.

To use: Apply a small amount to dry skin and massage in for 2 minutes. Tissue off. Rinse with warm water if desired.

111

Lavender & Rosemary Purifying Cleanser

Soap-based cleanser
Makes approx. 75 ml
Best for normal, oily, combination or blemish-prone skin

Rosemary essential oil is antiseptic and astringent, which makes it a very good choice for blemish-prone skin.

3 tablespoons boiling water
2 teaspoons plain table salt
4 tablespoons lavender liquid
  Castile soap
10 drops rosemary essential oil

1. Pour the boiling water into a small glass measuring jug. Add the salt and stir well to dissolve. Set aside.
2. Add the lavender liquid Castile soap and essential oil to a sanitised bottle with a lid or pump.
3. Place a funnel in the top of the bottle and add 1 tablespoon of the salt solution to the liquid Castile soap. Discard any remaining salt solution.
4. Put the lid on the bottle and shake well to thicken. Store in the fridge and use within 2 weeks.

To use: Shake the bottle well and apply a pea-sized amount to wet skin, massage in (avoiding your eyes) and rinse off with warm water.

112

Argan Oil & Oat Facial Cleanser

Soap-based cleanser
Makes approx. 100 ml
Best for all skin types

Argan oil is rich in vitamin E, phenols, carotenes and essential fatty acids, making it the perfect oil for all skin types. The Helichrysum essential oil has anti-inflammatory properties to help soothe and protect the skin.

2 tablespoons liquid Castile soap
2 tablespoons Aloe vera gel
1 tablespoon finely ground oats
1 tablespoon argan oil
1 tablespoon vegetable glycerin
5 drops rose essential oil
5 drops Helichrysum essential oil

1. Combine all of the ingredients in a sanitised bottle. Shake the bottle well. Store in the fridge and use within 2 weeks.

To use: Shake the bottle well and apply a generous amount to wet skin, massage in (avoiding your eyes) and rinse off with warm water.

▲ Lemon verbena

113

Lemon Verbena Cleansing Milk

Water-based cleanser
Makes approx. 120 ml
Best for oily, normal or combination skin

Lemon verbena floral water is softly stimulating to the senses, with astringent and cleansing properties for the skin.

120 ml lemon verbena floral water
2 tablespoons whole milk
1 tablespoon vegetable glycerin
$\frac{1}{4}$ teaspoon liquid lecithin
10 drops lemon essential oil

1. Combine the lemon verbena floral water, milk, vegetable glycerin and liquid lecithin in a small mixing bowl.
2. Using a handheld mixer, blend the ingredients together on medium speed until they are combined and have the consistency of a gel.
3. Add the essential oil and blend to combine. Store in the fridge and use within 1 week.

To use: Apply a generous amount to the skin, massage in with your fingertips and tissue or rinse off.

◀ Plain table salt, finely ground oats and rosemary

◄ Chamomile Cream &
Honey Dry Skin Cleanser

114

Chamomile Cream & Honey Dry Skin Cleanser

Water-based cleanser
Makes approx. 100 ml
Best for dry, sensitive or normal skin

Chamomile is a very soothing herb that is well known for its ability to calm the skin. You get a double dose of chamomile in this formula.

60 ml water
1 tablespoon dried chamomile flowers
2 tablespoons double cream
1 tablespoon honey
5 drops German chamomile essential oil

1. Bring the water to a boil in a small saucepan and add the dried chamomile flowers.
2. Turn off the heat and cover the saucepan with a lid until the liquid cools to room temperature.
3. Strain the liquid from the herb and pour into a sanitised bottle.
4. Add the cream, honey and essential oil. Shake well to combine. Store in the fridge and use within 1 week.

To use: Apply a generous amount to the skin, massage in with your fingertips and tissue or rinse off.

115

Thick & Rich Almond & Carrot Cleanser

Oil-based cleanser
Makes approx. 150 ml
Best for normal, dry or sensitive skin

The carrot seed essential oil in this cleanser is beneficial to mature skin and irritated skin, and can even help combat wrinkles.

60 ml sweet almond oil
2 teaspoons finely grated beeswax
¼ teaspoon liquid lecithin
2 teaspoons vegetable glycerin
70 ml rose water
15 drops carrot seed essential oil

1. Measure the sweet almond oil, beeswax, liquid lecithin and vegetable glycerin into a glass measuring jug that has been placed in a saucepan containing a few centimetres of simmering water.
2. Allow the wax to melt fully into the other ingredients.
3. Remove the jug from the heat and transfer the mixture to a heat-safe mixing bowl. Cool to 20–24°C (68–75°F).
4. Heat the rose water to a temperature of 20–24°C (68–75°F).
5. Once the oil and water phases are at the same temperature, begin beating the oil/wax/lecithin/glycerin mixture with a handheld mixer set on medium-high speed. Very slowly drizzle in the warm rose water. Continue mixing for about 5 minutes or until the liquid starts to thicken and emulsify.
6. Mix in the carrot seed essential oil. Transfer the mixture to a dry, sanitised bottle and allow to cool. Store in the fridge and use within 2 weeks.

To use: Apply a generous amount to the skin, massage in with your fingertips and tissue or rinse off.

116

Terrific Triple Oil Cleanser & Makeup Remover

Oil-based cleanser
Makes approx. 90 ml
Best for all skin types

Use this cleanser to deep cleanse and break up stubborn makeup while dissolving impurities.

2 tablespoons extra virgin olive oil
2 tablespoons hazelnut oil
2 tablespoons kukui nut oil
2 drops neroli essential oil
2 drops palmarosa essential oil
2 drops frankincense essential oil

1. Measure out all of the ingredients into a small, dry, sanitised bottle. Shake well to combine.

To use: Apply a small amount to dry skin and massage in for 2 minutes. Tissue off. Rinse with warm water if desired.

▷ Hazelnut oil

117

Witch Hazel & Green Tea Clay Cleanser

Water-based cleanser
Makes approx. 100 ml
Best for oily and blemish-prone skin

This is the perfect treatment to cleanse, decongest and remove stubborn sebum build-up. Powdered green tea (known as Matcha) has wonderful antioxidant properties.

60 ml witch hazel extract
2 tablespoons vegetable glycerin
1 tablespoon French green clay, plus more to thicken if needed
1 teaspoon powdered green tea (Matcha)
10 drops tea tree essential oil

1. Mix together the witch hazel extract and vegetable glycerin.
2. Sprinkle in the green clay and mix to combine.
3. Add the powdered green tea and essential oil – a thick paste should form; if it is too thin, add a small amount of clay. Store in the fridge and use within 2 weeks.

To use: Massage a small amount onto a moist face. Leave on for 2 minutes and rinse off with warm water.

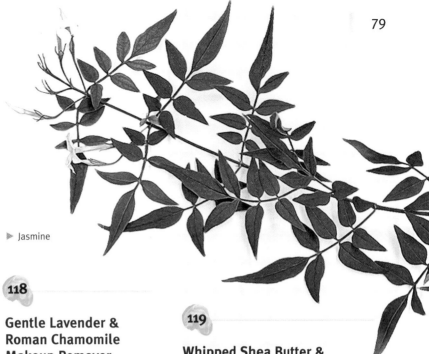

▶ Jasmine

118

Gentle Lavender & Roman Chamomile Makeup Remover

Oil-based cleanser
Makes approx. 115 ml
Good for all skin types

Gently cleanse and remove makeup while leaving your skin comforted and moisturised.

70 g palm shortening or coconut oil
2 tablespoons jojoba oil
2 drops lavender essential oil
2 drops Roman chamomile essential oil

1. Combine all of the ingredients in a small bowl. With a fork, whip the ingredients to combine well. Transfer the mixture to a dry, sanitised container.

To use: Apply a small amount to the face and massage in with your fingertips. Tissue off.

119

Whipped Shea Butter & Jasmine Clay Cleanser

Oil-based cleanser
Makes approx. 115 ml
Good for dry skin

Jasmine absolute oil is a treat for the senses. It smells truly amazing! Dry skin will be left highly hydrated and soft.

110 g shea butter
2 teaspoons rhassoul clay
5 drops jasmine absolute oil

1. Put the shea butter in a small glass measuring jug and place in a saucepan containing a few centimetres of simmering water. Allow the shea butter to melt completely.
2. Sprinkle the rhassoul clay over the butter mixture and whisk in to combine.
3. Place the jug in the fridge for 30 minutes to cool.
4. Blend the mixture with a whisk attachment for 10 minutes.
5. Add the jasmine absolute oil. Continue mixing for about 10 more minutes or until the shea butter has the texture of stiff whipped cream. Transfer to a dry, sanitised container.

To use: Massage a small amount over dry skin. Tissue off.

Note: Keep this mixture below 24°C (75°F) or it may melt.

Facial Scrubs and Exfoliators

Exfoliators are skin-smoothing and perfecting treatments for the skin, prepared with gentle and natural polishing ingredients such as sweet sugar, finely ground salt, powdered potent herbs, ground nuts and pulverised apricot kernels. Scrubs and exfoliators can polish away lifeless skin cells in a soothing way to uncover a radiant, healthy-looking skin tone in a few minutes.

Facial scrubs can be sugar, salt, bicarbonate of soda or flower and herb based.

TRY IT

A simple paste of bicarbonate of soda and warm water costs a few pennies but can do wonders – smoothing the skin and erasing dry patches.

120

Choosing and using different types of scrub and exfoliator

Sugar scrubs: Use caster sugar as the exfoliator. Usually combined with a nourishing oil, these wonderful formulas can help restore a radiant complexion with visibly improved skin texture and clarity for all skin types.

Salt scrubs: Use extra-fine sea salt, such as sel gris, along with a pampering carrier oil to exfoliate the skin. Best used on normal, oily and blemish-prone skin types.

Herbal and flower scrubs: Use powdered herbs and flowers such as ground rose petals, maize flour, powdered grapefruit peel and even powdered thyme to gently polish and energise the skin. Wonderful for all skin types, especially dry, sensitive and mature skins.

Bicarbonate of soda, clay and other types of scrub: These may combine natural ingredients such as bicarbonate of soda, clay, ground almonds, pulverised walnut shells and apricot kernel meal to polish away dull, dry skin and promote circulation. When combined with a carrier oil or butter, they will also hydrate and pamper the skin.

121

The benefits of using facial scrubs

Using facial scrubs on a regular basis will have the following benefits:

- Removes dead skin cells.
- Combats blemishes.
- Reduces discolourations.
- Softens the skin.
- Polishes away uneven dry patches for a perfect, even-toned complexion.
- Helps reduce the appearance of pores, fine lines and wrinkles.

122

Extra hydration

When deciding on which liquid to use to make a paste from a dry scrub, you can substitute milk for water if your skin needs extra hydration.

123

How to use a facial scrub

Try to make using a facial scrub a regular part of your weekly skincare routine.

1. Moisten clean skin with a little warm water.
2. Apply a small amount of exfoliator to the skin, and gently and slowly massage in circular motions, avoiding the eye area, for 1–2 minutes.
3. Leave the scrub on for 3–5 minutes before rinsing away with warm water and gently patting the face dry.

124

Slippery scrubs

When using a scrub that contains oil in the bath or shower, take special care not to slip because the floor may become very slippery.

125

How often to use a scrub

Care needs to be taken when applying a scrub to your face. Only use a face scrub periodically and with light pressure against your skin. Follow this chart for best results.

Skin Type	Weekly Facial Scrubs
Normal skin	Two to three times per week
Oily/blemish-prone skin	Two to three times per week
Combination skin	Once or twice a week
Sensitive skin	Once a week
Mature and dry skin	Once or twice a week, with three days between each treatment

FIX IT

The introduction of any water or moisture into the container can cause mould to begin to grow, which will ruin your product. To combat this, always use a clean, dry spoon to retrieve your scrub formula.

Recipes:
Six of the Best Facial Scrubs and Exfoliators

126

Lovely Lavender Purifying Face Scrub

Makes enough for 2 treatments
Best for all skin types

Brighten your complexion by gently exfoliating away dry, lifeless skin with this lovely scrub.

2 teaspoons rhassoul clay
2 teaspoons finely ground oats
1 teaspoon almond meal
1 teaspoon maize flour
1 teaspoon lavender buds
¼ teaspoon powdered neem leaf
Warm water (enough to make a paste)
10 drops lavender essential oil

1. In a small bowl, combine the rhassoul clay, oats, almond meal, maize flour, lavender buds and neem leaf. Stir to combine.
2. Add warm water in teaspoon increments until a smooth paste forms.
3. Stir in the lavender essential oil. Store in a tightly lidded container and use within 7 days.

To use: Moisten your face and apply half of the scrub. Using a gentle pressure, massage the skin with circular motions for 1–2 minutes. Leave the mask on for 5 minutes and rinse away with warm water.

127

Oh Yes! Oaty Facial Scrub

Makes enough for 2–3 treatments
Best for all skin types

This scrub helps unveil a fresher and more even-toned skin.

1 tablespoon sweet almond oil
1 tablespoon hemp seed oil
1 tablespoon vegetable glycerin
1 tablespoon finely ground
  oats
1 tablespoon caster sugar
5 drops ylang ylang essential oil

1. Combine all of the ingredients in a small bowl and mix well to combine. Store in a tightly lidded container.

To use: Moisten clean skin with some warm water. Apply a generous amount of scrub to the palm of your hand and add a tiny amount of warm water to form a moist paste. Massage onto the skin (avoiding the eye area) with gentle circular motions for 1–2 minutes. Leave on the skin for 5 minutes and rinse away with warm water.

128

Brilliant Blueberry & Manuka Honey Face Scrub

Makes enough for 5 treatments
Best for all skin types

A sweet, age-smoothing face scrub, which is made with antioxidant-rich blueberries and skin-smoothing manuka honey. Reveal your luminous skin!

1 tablespoon ground freeze-dried blueberries
4 tablespoons caster sugar
2 teaspoons Kaolin clay
½ teaspoon ground cinnamon
2 tablespoons sweet almond oil
2 tablespoons castor oil
1 tablespoon manuka honey
20 drops carrot seed essential oil
5 drops Frankincense essential oil

1. Combine all of the ingredients in a small bowl and mix well to combine. Store in a tightly lidded container and use within 7 days.

To use: Moisten clean skin with some warm water. Apply a generous amount of scrub to the palm of your hand and add a tiny amount of warm water to form a moist paste. Massage onto the skin (avoiding the eye area) with gentle circular motions for 1–2 minutes. Leave on the skin for 5 minutes and rinse away with warm water.

▶ Brilliant Blueberry & Manuka Honey Face Scrub

129

Lemon & Lavender Salt Scrub for the Face

Makes enough for 2–3 treatments
Best for normal, oily, blemish-prone and combination skin

This is a very invigorating scrub that removes sebum build-up and loosens pore-clogging impurities.

1 tablespoon hazelnut oil
1 tablespoon jojoba oil
2 tablespoons extra-fine sea salt
2 teaspoons finely grated fresh lemon zest
2 teaspoons finely ground dried lavender buds
1 teaspoon Kaolin clay
10 drops lavender essential oil
5 drops lemon essential oil

1. Combine all of the ingredients in a small bowl and mix well to combine. Store in a tightly lidded container.

To use: Moisten clean skin with some warm water. Apply a generous amount of scrub to the palm of your hand and add a tiny amount of warm water to form a moist paste. Massage onto the skin (avoiding the eye area) with gentle circular motions for 1–2 minutes. Leave on the skin for 5 minutes and rinse away with warm water.

▼ Lemon & Lavender Salt
Scrub For The Face

130

Tea Tree, Thyme & Bicarbonate of Soda Facial Scrub

Makes enough for 2 treatments
Best for oily and blemish-prone skin

This is a blemish-reducing and detoxifying facial mask for those with oily or blemish-prone skin.

2 tablespoons bicarbonate of soda
Warm water (enough to make a paste)
1 teaspoon jojoba oil
6 drops tea tree essential oil
5 drops thyme (ct. linalool) essential oil

1. Combine the bicarbonate of soda and warm water to form a paste.
2. Stir in the jojoba oil, as well as the tea tree and thyme essential oils. Store in a tightly lidded container and use within 7 days.

To use: Moisten clean skin with some warm water. Apply a generous amount of scrub to the face. Massage onto the skin (avoiding the eye area) with gentle circular motions for 1–2 minutes. Leave on the skin for 5 minutes and rinse away with warm water.

131

Radiance-Revealing Rose Petal & Grapefruit Peel Facial Scrub

Makes enough for 2 treatments
Best for all skin types

A very gentle, yet potent, skin scrub, which is suitable for every skin type.

1 teaspoon powdered dried rose petals
1 teaspoon powdered dried grapefruit peel
1 teaspoon rose water
1 teaspoon vegetable glycerin
2 drops rose essential oil (optional)

1. Combine all of the ingredients in a small bowl and mix well to combine. Store in a tightly lidded container and use within 7 days.

To use: Moisten clean skin with some warm water. Apply a generous amount of scrub to the palm of your hand and add a tiny amount of warm water to form a moist paste. Massage onto the skin (avoiding the eye area) with gentle circular motions for 1–2 minutes. Leave on the skin for 5 minutes and rinse away with warm water.

Face Masks

Face masks are extraordinary treatments for the face and offer numerous benefits depending on the precise requirements of the skin. For example, face masks can help purify blemish-prone skin, tighten pores on oily skin, soothe sensitive skin, revitalise dry skin and even help revivify aged skin. Many face masks are formulated with clays and oils, and are worn on the skin for up to one hour to achieve their remarkable effects.

FIX IT

Clumpy face mask? Try using a fork instead of a spoon when mixing a clay-based mask to avoid lumps.

 132

Choose the right face mask for your skin type

Masks for dull skin: These are the perfect pick-me-up for tired and dull-looking skin, and are often formulated with powerful, skin-nourishing ingredients such as honey, shea butter, cocoa butter, rosehip seed oil, Helichrysum essential oil, rose essential oil, as well as gentle clays, like rhassoul and kaolin, Aloe vera gel and floral waters.

Masks for oily and acne-prone skin: These are formulated with deep-cleansing ingredients such as Fuller's Earth clay, bentonite clay, activated charcoal, neem leaf powder, peppermint essential oil, tea tree essential oil, thyme tincture and witch hazel extract.

Masks for dry skin: Very hydrating and soothing for those with sensitive, dry or mature skin types, these masks are often made with rich butters, nourishing oils, gentle clays, chamomile essential oil, lavender essential oil, honey, yoghurt, fresh cream and rolled oats.

 133

Amazing clays

Use the absorbing power of clay to soak up excess oil and sebum, leaving the skin smoother, softer and brighter. Fresh-made clay face masks help to temporarily reduce the appearance of pores. They can be worn for up to 1 hour or until dry. They can also be mixed with ingredients such as honey, yoghurt, floral waters, teas and even fruit juices.

134

How to properly apply a face mask

A properly applied face mask will be in a generous and even layer over your face. For best results, wrap your hair in a towel or pull it away from your face and into a ponytail. Avoid getting the mask into your eyes, mouth and nose.

1. Cleanse the skin and gently pat dry.
2. Apply a generous amount of the mask to your face and neck (avoiding the eye area).
3. Leave on for 10 minutes to 1 hour.
4. Rinse off with some warm water.
5. Gently pat your face dry with a towel.

135

Delicate skin

Face masks are not meant to be applied to the delicate area around the eyes. Steer clear of this area to avoid drying out the skin.

TRY IT

• If you want to clear up a blemish fast, spot-treat it with a dab of clay-based face mask. Allow to air-dry and rinse away with warm water.

• If you are in a hurry, and don't have time to allow your mask to air-dry, turn your hairdryer to the cool setting and blast it dry in just a few minutes.

Recipes:
Six of the Best Face Masks

 136

Fresh Oat & Banana Face Mask

Makes enough for 2 masks
Best for sensitive, dry or mature skin

This is a moisturising and soothing face mask for sensitive skin. The addition of the glycerin helps your skin retain important moisture.

2 tablespoons finely ground oats
2 tablespoons ground almond meal
1 medium banana, peeled
1 tablespoon glycerin
2 teaspoons jojoba oil
20 drops vitamin E

1. Combine all of the ingredients in a food-processor and process to a smooth paste. Store in the fridge and use within 5 days.

To use: Apply half the mask to clean skin (avoiding the eye area). Leave on for 15–20 minutes and then rinse off with warm water.

 137

Go Away Blemishes! Garlic & Rose Facial Mask

Makes enough for 3 masks
Best for oily and blemish-prone skin

This mask may not smell as pretty as some, but the potent garlic, astringent witch hazel and blemish-blasting thyme will work wonders on blemish-prone skin!

20 g Fuller's Earth clay
1 tablespoon witch hazel extract
1 teaspoon thyme tincture
1 small egg yolk
2 teaspoons honey
½ teaspoon neem-infused oil
2 large fresh garlic cloves
20 drops tea tree essential oil
5 drops rose essential oil

1. Combine all of the ingredients in a food-processor and process to a smooth paste. Store in the fridge and use within 5 days.

To use: Apply a generous amount to clean skin (avoiding the eye area). Leave on for 15–20 minutes and then rinse off with warm water.

 138

Chocolate-Covered Strawberry Face Mask

Makes enough for 2 masks
Best for all skin types

This delicious face mask is the perfect treat when you are craving chocolate but don't want those calories. This makes enough for you and someone special to enjoy the mask together.

4 large fresh strawberries
2 teaspoons cocoa powder
2 teaspoons rhassoul clay
½ teaspoon crushed dried rose petals
1 tablespoon Aloe vera gel
1 teaspoon vegetable glycerin
10 drops vanilla absolute

1. Combine all of the ingredients in a food-processor and process to a smooth paste. Store in the fridge and use within 5 days.

To use: Apply half the mask to clean skin (avoiding the eye area). Leave on for 15–20 minutes and then rinse off with warm water.

▲ Fresh Oat & Banana Face Mask

▲ Chocolate-Covered Strawberry Face Mask

139

Creamy Complexion Milk Mask

Makes enough for 2 masks
Good for normal, dry, sensitive and mature skin

This wonderful mask will leave your complexion feeling so smooth, nourished and glowing.

60 ml double cream
1 teaspoon ground oats
2 teaspoons kaolin clay
5 drops palmarosa essential oil

1. Measure out all of the ingredients into a small mixing bowl and mix to combine. Store in the fridge and use within 5 days.

To use: Apply half the mask to clean skin (avoiding the eye area). Leave on for 15–20 minutes and then rinse off with warm water.

140

Greek Yoghurt & Sea Kelp Pore-Cleaning Clay Mask

Makes 1 mask
Good for all skin types

The Greek yoghurt will soften your skin while the clay, kelp and Spirulina powders purify your pores.

1 tablespoon plain Greek yoghurt
¼ teaspoon kelp powder
¼ teaspoon Spirulina powder
¼ teaspoon French green clay

1. Mix all of the ingredients together in a small bowl to form a paste.

To use: Apply to clean skin (avoiding the eye area). Leave on for 15–20 minutes and then rinse off with warm water.

141

Radiant Skin Clay & Activated Charcoal Face Mask

Makes enough for 2 masks
Good for normal, combination, blemish-prone and oily skin

This mask is a very potent skin-cleansing formula that will leave your face looking and feeling super purified, healthy and radiant.

1 tablespoon French green clay
½ teaspoon activated charcoal
½ teaspoon green tea (Matcha) powder
1 tablespoon Aloe vera gel
2 teaspoons witch hazel extract, plus more if needed to thin the mixture
1 teaspoon vegetable glycerin
10 drops eucalyptus essential oil
5 drops rosewood essential oil
8 drops cypress essential oil

1. Mix all of the ingredients together to form a paste (adding more witch hazel extract in ¼-teaspoon increments to thin the mixture if it is too thick). Store in the fridge and use within 5 days.

To use: Apply half the mask to clean skin (avoiding the eye area). Leave on for 15–20 minutes and then rinse off with warm water.

◀ Creamy Complexion Milk Mask

Facial Steams

Facial steams use herb- and flower-infused steam to open up the pores in your skin in order to purify, soften and hydrate, as well as to increase circulation. Your dull complexion is left glowing and purified after a herbal steam treatment. Facial steams are suitable for normal, oily, combination, dry and mature skin types. However, those with sensitive, blemish-prone or irritated skins should not use a facial steam.

TRY IT

Instead of tossing out any leftover herb infusion, use it as a hair rinse (see pages 68–71).

142

How to make and use a facial steam

Facial steams should only be used up to once a week. Those with medical problems, such as diabetes and eczema, or eye problems, for example, should consult a healthcare professional before using a steam mask.

1. Bring 2 litres of water to a boil in a saucepan.
2. Add 4–6 tablespoons of herb/flower blend to the boiling water. (See the chart opposite for some skin-specific recipe blends.) Cover the saucepan with a lid and steep for 10 minutes.
3. Pour the herbs/flowers and hot water into a large heatproof bowl that has already been placed on the table where you are planning to enjoy the steam.
4. Place your face about 20–25 cm above the bowl and drape a large towel or pillowcase over your head in order to create a steam tent.
5. Enjoy the steam for up to 10 minutes. (Stop immediately if you feel too hot or uncomfortable.)

6. Strain the liquid from the spent herb, splash over your face and then pat dry.
7. Any remaining liquid can be stored in the fridge for up to 3 days and used as a toner or refresher for the skin.

TRY IT

Package up small bags of dried herbs, along with a printout of the directions, for a special gift for friends and family.

143

The benefits of using facial steams

Using facial steams regularly can have the following benefits for your skin:
- Opens up the pores in the skin.
- Hydrates the skin.
- Increases circulation.
- The treatment is relaxing.
- Helps to shed dead skin cells.

144

Trust your skin

Listen to your body: if your face feels too hot or uncomfortable, then STOP the steam treatment immediately and rinse your skin with cool water.

145

Time-saving tip

Want to do a facial steam in a hurry? Freeze the liquid herb infusion in an ice-cube tray. When you want to use them, simply heat the ice cubes in a microwave-safe bowl until they boil. Carefully follow the directions for the facial steam method.

146

Herb and flower recipe blends for herbal steams

Revitalise your specific skin-type with one of these pore-opening and rejuvenating facial steam recipes.

Skin Type	Recipe Blend (Makes enough for one facial steam)
Oily skin	1 tablespoon dried basil 1 tablespoon dried lemon balm 1 tablespoon dried witch hazel 1 tablespoon dried sage 1 tablespoon dried lavender
Congested skin and skins prone to blackheads	1 tablespoon dried elder flowers 1 tablespoon dried neem leaf 2 tablespoons dried thyme 1 teaspoon dried goldenseal 1 teaspoon dried Oregon grape root 1 tablespoon dried lavender
Normal and combination skin	1 tablespoon dried rose petals 1 tablespoon dried calendula flowers 1 tablespoon dried chamomile flowers 1 tablespoon dried lemon balm 1 tablespoon dried lavender 1 teaspoon dried peppermint
Dry and mature skin	1 tablespoon dried rose petals 1 tablespoon dried chamomile flowers 1 teaspoon dried peppermint 1 teaspoon dried fennel seed 1 tablespoon dried elder flowers 2 teaspoons dried Helichrysum flowers

▶ Chamomile flowers, lavender buds and rose petals are fantastic for combination skin.

Facial Toners

Facial toners are used on the skin after cleansing, exfoliating and the application of masking treatments in order to balance, soothe and purify the skin. Toners are put on just before serums and moisturisers, using a saturated cotton pad or a spray bottle. Toners may be formulated with floral waters, herbal infusions, apple cider vinegar and even vodka. They can also be spritzed on several times throughout the day for a refreshing boost or to gently hydrate the skin.

▼ Orange blossom

147

Choosing and using different facial toners

Floral-water toners: These are very gentle, floral-water-based toners that are wonderful for sensitive, dry and mature skin types. The main ingredient is a floral water such as rose water or orange blossom water.

Herbal-infused toners: Witch hazel is a popular herb-infused toner that works as an astringent for the skin. Herbal-infused toners are easy to make, and fresh herbs, dried herbs and even tinctures can be used to make skin-specific formulas.

Vinegar toners: These are pH-balancing toners that can be used on all skin types.

TRY IT

To help remove dead surface skin cells, apply the toner to a clean, soft flannel and massage the skin in gentle circular motions.

148

No peeking!

Ensure that your eyes are tightly closed before misting on a facial toner. Vinegar-based and astringent toners can burn and sting if introduced into the eyes.

149

Chill out

Store your toner in the fridge on hot, sunny days. Then mist your skin with a cool and refreshing toner to soothe and balance.

FIX IT

Make sure that you shake toners well before use to distribute all of the ingredients evenly.

◀ Witch hazel

Recipes:
Three of the Best Facial Toners

150

Soothing Rose Geranium & Chamomile Toner

Makes approx. 120 ml
Best for all skin types

This lovely toner hydrates and helps plump up your skin.

60 ml rose geranium floral water
2 tablespoons dried chamomile
* flowers*
2 tablespoons Aloe vera gel
2 tablespoons vodka
20 drops Roman chamomile
* essential oil*
8 drops rose geranium essential oil

1. Pour the rose geranium floral water into a small saucepan over a medium-low heat. Bring to a simmer.
2. Turn off the heat and add the dried chamomile flowers. Cover with a lid and cool to room temperature.
3. Strain the liquid from the spent herb mix.
4. Transfer to a sanitised bottle with a lid.
5. Add the Aloe vera gel, vodka and essential oils. Shake well. For best results, store in a cool, dark place and use within 60 days.

To use: Apply the toner to your cleansed face and neck area with clean cotton pads, avoiding the eye area.

151

Toner for Blemish-Prone Skin

Makes approx. 120 ml
Best for oily, blemish-prone and normal skin

A purifying and astringent toner to help clear blemish-prone skin.

1 tablespoon lavender floral water
1 tablespoon rose water
3 tablespoons witch hazel extract
1 tablespoon Aloe vera gel
1 tablespoon calendula tincture
1 tablespoon thyme tincture
10 drops palmarosa essential oil
10 drops carrot seed essential oil
5 drops myrrh essential oil
5 drops German chamomile essential oil

1. Add all of the ingredients to a sanitised bottle with a lid. Shake well. For best results, store in a cool, dark place and use within 60 days.

To use: Apply the toner to your cleansed face and neck area with clean cotton pads, avoiding the eye area.

152

Toner for Sensitive or Dry Skin

Makes approx. 120 ml
Best for sensitive and dry skin

This is a potent toner made with a wonderful variety of floral waters to help balance the skin and improve dry and sensitive skin.

2 tablespoons rose water
1 tablespoon calendula floral water
1 tablespoon orange blossom water
1 tablespoon chamomile floral water
1 tablespoon lavender floral water
2 teaspoons apple cider vinegar
2 teaspoons vegetable glycerin
5 drops rosewood essential oil
2 drops rose essential oil
2 drops neroli essential oil
2 drops Roman chamomile essential oil

1. Add all of the ingredients to a sanitised bottle with a lid. Shake well. For best results, store in a cool, dark place and use within 60 days.

To use: Apply the toner to your cleansed face and neck area with clean cotton pads, avoiding the eye area.

Facial Moisturisers

Every skin type needs a good facial moisturiser. A natural plant-based moisturiser makes your skin feel amazingly soft and supple, plus it delivers precious protection from the outside elements.

Do you want to create a rich, thick and concentrated nourishing face cream for your dry and parched skin? Perhaps you have blemish-prone skin that requires a lightweight facial lotion made with skin soothing and healing essential oils. How does a quickly absorbing beauty oil that leaves your skin silky and supple sound?

It is easy to customise your ideal facial moisturiser by simply blending together your favourite nourishing plant oils, butters, essential oils and herbal extracts for a healthy and noticeably glowing face and neck.

TRY IT

If you have very dry or mature skin, massage two to three drops of facial oil onto your face and follow with a rich face cream for extra hydration and protection.

153

The best way to apply moisturiser

The goal to applying any moisturiser is to keep your skin hydrated and protected. After you have cleansed and toned your skin, apply small pea-sized dabs of your moisturiser to your forehead, cheeks, nose, chin and neck. Gently massage the moisturiser into your face and neck using small and gentle circular strokes with your fingertips. Continue to gently massage for about a minute or until the moisturiser is fully absorbed. Use moisturisers sparingly for daytime use and more liberally for night time use.

154

Make an emulsion

If you suffer from dry skin, an emulsion-based moisturiser is great for hydrating your skin. If you have ever made mayonnaise at home, you understand the emulsion process. An emulsion is a permanent and stable combination of oil and water, which normally do not mix well together. Beeswax and emulsifying wax are the most common ingredients used when making creams and lotions. They work to create a lasting bond between the oil and water so the final product is stable and does not separate. Stearic acid, liquid lecithin, lanolin, xanthan gum and guar gum powder are commonly used to thicken and emulsify natural beauty products.

155

Choosing different types of moisturiser

There are two broad categories of moisturisers: oil-based formulas and oil- and water-based formulas. Within each category there are specific types of moisturisers. See the chart below, which illustrates the types of facial moisturisers and the skin types they suit.

TRY IT

Massage a few drops of the Nourishing Facial Oil (see page 94) into your dry, brittle cuticles before bedtime and wake up with hydrated, healthy nails.

Oil-Based Formulas	Oil- and Water-Based Formulas
Face Balms: Formulated with carrier oils, butters, waxes and essential oils. Can be used over the entire face and offers superior protection and hydration. **Best for** mature, dry, normal and sensitive skin.	*Face Lotions:* Formulated by emulsifying a water phase and an oil phase together. Lotions are not as thick as creams and are more easily absorbed by the skin. Lotions have a higher ratio of water to oil ingredients. **Best for** normal, combination, oily and blemish-prone skin.
Face Salves: Formulated with herb-infused carrier oils, waxes and essential oils. Rich formulas that offer immediate protection and deep hydration. Often used to soothe and relieve mild and irritating skin conditions. **Best for** mature, dry, normal and sensitive skin.	*Face Creams:* Formulated by emulsifying a water phase and an oil phase together. Creams are thicker than lotions and usually contain a higher percentage of oil to water. **Best for** dry, mature, normal and sensitive skin.
Facial Oils: Formulated with herb-infused carrier oils, plain carrier oils and essential oils. A little goes a long way and should be used sparingly. **Best for** all skin types to balance, calm, hydrate and smooth.	

156

Make the best basic beeswax beauty cream

Ingredients:
• *65 ml olive oil*
• *2 teaspoons beeswax, finely grated*
• *1 teaspoon liquid lecithin*
• *60 ml distilled water*
Makes approx. 150 ml

1. Measure the olive oil, grated beeswax and liquid lecithin into a glass measuring jug that has been placed in a saucepan with a few centimetres of simmering water. Allow the wax to fully melt into the olive oil and lecithin. Remove from the heat and transfer the mixture to a heat-safe mixing bowl to cool to between 20 and 24°C (68 and 75°F).
2. Heat the distilled water to between 20 and 23°C (68 and 73°F).
3. Once both phases are the same temperature, begin beating the oil/wax lecithin mixture with a handheld mixer set on medium-high speed. Very slowly drizzle in the warm distilled water. Continue mixing for about 5 minutes or until the liquid starts to thicken and emulsify.
4. Transfer to a sterile container and allow to cool. Store in the fridge and use within 2 weeks.

◀ Remember: beeswax must be finely grated to enable it to melt easily.

TRY IT

Most handcrafted face products are multi-purpose. Try using:
• Facial oil as a cuticle oil
• Face salve to smooth split-ends and control fly-aways in your hair
• Face cream as a soothing foot cream
• Facial lotion as an ultra hydrating body lotion

Recipes:
Five of the Best Facial Moisturisers

▶ Shea butter

157

Nourishing Facial Oil

Makes approx. 60 ml
Best for dry, mature, normal and combination skin

A very luxurious recipe to soothe and soften your face with a blend of hydrating rose hip seed oil, jojoba oil and skin-rejuvenating carrot seed essential oil.

2 tablespoons jojoba oil
1 tablespoon plus 1 teaspoon rose
  hip seed oil
1¼ teaspoons vitamin E
7 drops carrot seed essential oil
2 drops rose essential oil
3 drops Roman chamomile
  essential oil

1. Mix all the ingredients into a small amber-coloured dropper bottle. Shake well before use.

To use: Apply up to five drops to a cleansed and toned face.

158

Intensive Night Time Facial Cream

Makes approx. 60 ml
Best for dry, mature, normal, combination and sensitive skin

This is a very thick and rich facial cream that combines three moisturising carrier oils, ultra-rich shea butter and vitamin E together, along with the essential oils of sandalwood and rosewood to nourish and moisturise your skin while you sleep.

2 tablespoons orange blossom water
½ teaspoon vegetable glycerin
¾ teaspoon avocado oil
¾ teaspoon olive oil
½ teaspoon jojoba oil
1½ teaspoons shea butter
½ teaspoon stearic acid
1 heaping teaspoon emulsifying wax NF
⅛ teaspoon vitamin E
8 drops sandalwood essential oil
5 drops rosewood essential oil

1. Measure out the orange blossom water and vegetable glycerin into a glass measuring jug and sit it in a saucepan with a few centimetres of simmering water.
2. Measure out the avocado oil, olive oil, jojoba oil, shea butter, stearic acid and emulsifying wax into a glass measuring jug and sit it in a saucepan with a few centimetres of simmering water.
3. When both mixtures have reached 66°C (150°F), remove from the heat.
4. Carefully pour the oil mixture into a heat-proof mixing bowl and begin mixing with a hand mixer set on medium speed. Add in the orange blossom and glycerin and continue to mix for 5 minutes.
5. Once the mixture cools to under 38°C (100°F), mix in the vitamin E, and the sandalwood and rosewood essential oils. Pour your cream into a sterile container and allow to cool completely. Store in the fridge and use within 2 weeks.

To use: After cleansing and toning your face, apply a generous amount to your face and neck before bed, and gently massage in using small, circular motions.

▶ Beeswax

159

Tea Tree Oil & Cucumber Balancing Day Lotion

Makes approx. 60 ml
Best for normal, combination, oily and acne-prone skin

A light, refreshing and purifying lotion that is perfect for oily and blemish-prone skin. Made with fast-absorbing jojoba and hazelnut oils along with skin-toning witch hazel and purifying tea tree essential oil.

2 tablespoons cucumber hydrosol
2 teaspoons witch hazel extract
1¼ teaspoons jojoba oil
1 teaspoon hazelnut oil
1 heaping teaspoon emulsifying wax NF
½ teaspoon stearic acid
⅛ teaspoon vitamin E
10 drops tea tree essential oil

1. Measure out the cucumber hydrosol and witch hazel into a glass measuring jug and sit it in a saucepan with a few centimetres of simmering water.
2. Measure out the jojoba oil, hazelnut oil, stearic acid and emulsifying wax into a glass measuring jug and sit it in a saucepan with a few centimetres of simmering water.
3. When both mixtures have reached 66°C (150°F), remove from the heat.
4. Pour the oil mixture into a heat-proof mixing bowl and begin mixing with a hand mixer set on medium speed. Carefully add in the warm cucumber hydrosol and witch hazel and continue to mix for 5 minutes.
5. Once the mixture cools to under 38°C (100°F), mix in the vitamin E and tea tree essential oil.
6. Pour your cream into a sterile container and allow to cool completely. Store in the fridge and use within 2 weeks.

To use: After cleansing and toning, apply small pea-sized dabs of your moisturiser to your face and neck, and gently massage until the moisturiser is fully absorbed.

160

Rejuvenating Argan Oil & Rose Facial Balm

Makes approx. 60 ml
Best for all skin types

Great for all skin types. This hydrating facial balm quickly absorbs into the skin and leaves it smooth, silky and hydrated.

2 teaspoons grapeseed oil
2 teaspoons shea butter
2 teaspoons mango butter
1½ teaspoons grated beeswax
4 teaspoons argan oil
½ teaspoon vitamin E oil
7 drops rose essential oil

1. Measure out the grapeseed oil, shea butter, mango butter and beeswax into a small glass measuring jug and sit it in a saucepan with a few centimetres of simmering water until fully melted. Turn off the heat but leave the measuring jug sitting in the hot water.
2. Stir in the argan oil, vitamin E oil and rose essential oil. Quickly pour into your chosen heat-proof container and allow to cool completely.

To use: Apply small pea-sized dabs to a cleansed and toned skin. Gently massage into the face and neck using small circular motions.

161

All-Purpose Healing Face Salve

Makes approx. 60 ml
Best for dry, mature, normal and sensitive skin

Healing calendula flowers are infused into olive oil and thickened with beeswax to create a melt-into-your-skin salve that helps heal and soothe dry and parched skin.

2 tablespoons calendula-infused olive oil
2 teaspoons castor oil
2 heaping teaspoons grated beeswax
1 teaspoon vitamin E
5 drops carrot seed essential oil
5 drops Helichrysum essential oil

1. Measure out the calendula-infused olive oil, castor oil and beeswax into a small glass measuring jug and sit it in a saucepan with a few centimetres of simmering water until fully melted. Remove from the heat.
2. Add in the vitamin E and essential oils. Quickly pour into your chosen heat-proof container and allow to cool completely.

To use: Apply small pea-sized dabs to a cleansed and toned skin. Gently massage into the face and neck using small circular motions.

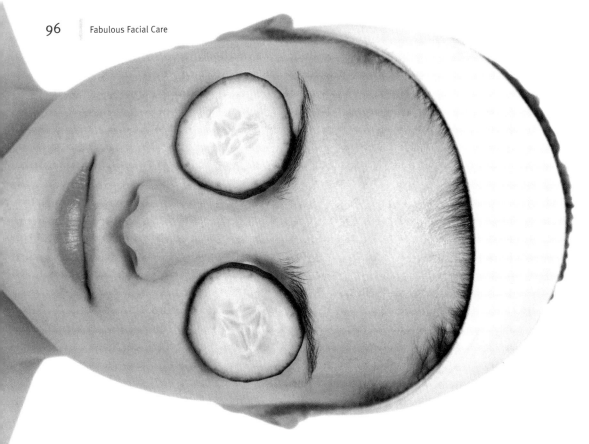

Eye Care

The tender skin near your eye is very delicate and thin. Care must be taken to nourish this area of the face with gentle botanicals and plant oils. Many natural ingredients work to help soften the appearance of dark circles and reduce puffiness. Harness the power of antioxidant-rich oils to hydrate the eye area, as well as reduce the appearance of fine lines and wrinkles.

TRY IT

Dip two clean cotton pads into some rose water and place in a small mesh bag. Freeze the bag for 10 minutes. Apply the frozen bag to closed eyes for 5 minutes to de-puff and soothe.

162

Choosing and using eye-care treatments

Eye creams: These are thick and rich, and made from an emulsion of floral waters, oils and waxes. Botanical ingredients, such as tinctures, extracts and essential oils, are often added to help moisturise, soothe, calm and firm, as well as to reduce the signs of aging around the delicate eye area.

Eye serums: These oil-based formulas are made with antioxidant-rich plant oils, which are easily absorbed into the skin. Essential oils are added to help de-puff the eyes, stimulate microcirculation and help prevent the signs of aging.

Eye balms: These waterless formulas contain rich plant oils, beeswax and essential oils. A small amount is gently dabbed around the eye area at night and before applying makeup in the morning. Eye balms help to protect the skin barrier and to reduce the appearance of fine lines and wrinkles. Essential oils are added in small amounts, at around 0.5%.

163

Tips for tackling dark circles and puffiness

Try the ideas listed below to banish puffy eyes and dark circles for good.

- Make a herbal infusion by combining 2 tablespoons of boiling water with a teaspoon each of dried chamomile flowers, fennel seeds and lavender buds. Once the liquid has cooled to room temperature, squeeze the liquid from the herbs and add an ice cube to chill. Dip a cotton pad into the chilled infusion and apply to the eye area for 10 minutes. Repeat if desired.
- Yes, chilled cucumber slices really do work to prevent puffiness! Apply them to the eye area for up to 10 minutes.
- Freeze a mixture of rose water and Aloe vera gel in one of the compartments of an ice-cube tray. Wrap the frozen ice cube in a muslin-cloth bag and gently apply to the puffy areas of the eyes for a few minutes.

FIX IT

Using your fingertips to handle a product can quickly introduce bacteria and germs to the product. To avoid this, use a clean cotton pad instead.

164

The best way to apply eye-care products

The skin near your eyes is one of the most fragile and sensitive areas of the body so care must be taken to avoid irritation.

1. Cleanse and tone your skin as required.
2. Using a clean cotton pad, retrieve a small amount of the product from the container and place on your ring ringer. Gently dab it onto the skin underneath and around your eyes.
3. Very gently apply the product with a light pressure around the eye area until it is absorbed.
4. Avoid getting any of the product in your eyes.

165

Beautiful brows

When applying an eye treatment, don't forget about your eyebrows. Just like your hair, they will also benefit from being conditioned.

166

Keeping it cool

For the longest shelf-life possible, store your products in the fridge. As an added bonus, applying a chilled product to the eye area can also help to soothe and de-puff.

► Dried chamomile flowers and fennel seeds are great in herbal infusions.

Recipes:
Six of the Best Eye-Care Products

167

Basic Eye Cream

Makes approx. 30 ml
Best for all skin types

This quickly absorbed formula contains moisturising ingredients such as olive oil and shea butter, meaning that it glides on easily to hydrate the delicate under-eye area.

1 tablespoon distilled water
2 teaspoons vegetable glycerin
³⁄₄ teaspoon extra virgin olive oil
1 teaspoon shea butter
1 teaspoon liquid lecithin
¹⁄₂ teaspoon emulsifying wax NF

1. Measure out the distilled water and vegetable glycerin into a glass measuring jug and place in a saucepan containing a few centimetres of simmering water.
2. Measure out the extra virgin olive oil, shea butter, liquid lecithin and emulsifying wax into another glass measuring jug and place in a second saucepan containing a few centimetres of simmering water.
3. When both mixtures reach a temperature of 71°C (160°F), remove them from the heat.
4. Carefully pour the oil mixture into a heatproof mixing bowl and begin mixing with a handheld mixer set on medium speed. Carefully add the distilled water and glycerin mixture, and continue to mix for a further 5 minutes.
5. Pour the eye cream into a small, sanitised container and allow to cool completely. Store in the fridge and use within 15 days.

To use: Apply to the eye area, following the steps on page 97.

168

Basic Eye Serum

Makes approx. 15 ml
Best for all skin types

Pomegranate seed oil is a really splendid and deeply penetrating oil that is amazing for the skin. It intensely nourishes the outer epidermal layer and delivers potent antioxidant benefits.

2 teaspoons pomegranate seed oil
1 teaspoon jojoba oil
2 drops Helichrysum essential oil

1. Place all of the ingredients into a small dropper bottle. Shake the bottle well.

To use: Apply to the eye area, following the steps on page 97.

169

Basic Eye Balm

Makes approx. 30 ml
Best for all skin types

A smoothing eye balm that can help reduce the visible signs of aging around the eyes. Calendula-infused oil is wonderful for dry and damaged skin.

5 teaspoons calendula-infused hemp seed oil
1 teaspoon finely grated beeswax
10 drops vitamin E oil
2 drops rose essential oil

1. Measure out the hemp seed oil and beeswax into a small glass measuring jug and place in a saucepan containing a few centimetres of simmering water until the wax has fully melted. Remove from the heat.
2. Add the vitamin E and rose essential oils. Quickly pour the balm into your chosen heatproof container and allow to cool completely. Use within 3 months.

To use: Apply to the eye area, following the steps on page 97.

170

Daily Revitalising Eye Cream

Makes approx. 30 ml
Best for all skin types

Daily use of this revitalising eye cream can help to visibly reduce the appearance of fine lines and wrinkles. This product is enhanced with ultra-rich avocado oil, which is high in the vitamins A, B1, B2, D and E.

1 tablespoon rose floral water
2 teaspoons vegetable glycerin
½ teaspoon avocado oil
½ teaspoon evening primrose oil
1 teaspoon cocoa butter
1 teaspoon liquid lecithin
½ teaspoon emulsifying wax NF

1. Measure out the rose floral water and vegetable glycerin into a glass measuring jug and place in a saucepan containing a few centimetres of simmering water.
2. Measure out the avocado oil, evening primrose oil, cocoa butter, liquid lecithin and emulsifying wax into another glass measuring jug and place in a second saucepan containing a few centimetres of simmering water.
3. When both mixtures reach a temperature of 71°C (160°F), remove them from the heat.
4. Carefully pour the oil mixture into a heatproof mixing bowl and begin mixing with a handheld mixer set on medium speed. Carefully add the distilled water and glycerin mixture, and continue to mix for a further 5 minutes.
5. Pour the eye cream into a small, sanitised container and allow to cool completely. Store in the fridge and use within 15 days.

To use: Apply to the eye area, following the steps on page 97.

171

Age-Defense Rich Eye Balm

Makes approx. 30 ml
Best for all skin types

The trio of plant-infused oils in this eye balm helps to repair, moisturise and protect the delicate eye area.

1 teaspoon calendula-infused carrier oil
1 teaspoon plantain-infused carrier oil
3 teaspoons marshmallow-infused carrier oil
1 teaspoon finely grated beeswax
10 drops vitamin E oil
2 drops rose essential oil
1 drop Helichrysum essential oil
1 drops carrot seed essential oil

1. Measure out the three carrier oils and beeswax into a small glass measuring jug and place in a saucepan containing a few centimetres of simmering water until the wax has fully melted. Remove from the heat.
2. Add the vitamin E oil and three essential oils. Quickly pour the balm into your chosen heatproof container and allow to cool completely. Use within 3 months.

To use: Apply to the eye area, following the steps on page 97.

172

Fortifying & Soothing Eye Serum

Makes approx. 15 ml
Best for all skin types

A wealth of luxurious plant oils are bottled up in this formula. Sea buckthorn oil alone is highly prized as an amazing oil for protecting the skin and helping to prevent wrinkles.

¼ teaspoon apricot kernel oil
¼ teaspoon castor seed oil
½ teaspoon jojoba seed oil
¾ teaspoon argan oil
¾ teaspoon pomegranate seed oil
⅛ teaspoon vitamin E oil
⅛ teaspoon sea buckthorn oil
2 drops Helichrysum essential oil
1 drop sandalwood essential oil
1 drop patchouli essential oil

1. Place all of the ingredients in a small dropper bottle. Shake the bottle well.

To use: Apply to the eye area, following the steps on page 97.

Lip Care

Pucker up and say farewell to dry, flaky and chapped lips – forever! Hydrate and protect your pretty pout with plant-based oils, butters and essential oils. Your lips will be softened, conditioned and naturally nourished.

173

Choosing and using lip-care products

Lip balms: These plant-oil and wax-based formulas can be used to heal, hydrate, protect and soften the lips to leave them feeling soft and smooth. Lip balms normally come in either a roll-up tube or a small pot. To use, simply apply to the lips as needed.

Tinted lip glosses: A little silkier than lip balms, these glosses glide on easily. They contain a natural tint from herbs such as alkanet root powder, beetroot powder, cocoa powder and hibiscus powder. Naturally tinted lip glosses are wonderful for soothing and protecting your lips, while making them sparkle subtly.

Lip scrubs: These are thick and rich exfoliating and conditioning treatments for your lips. Made from rich butters and gentle scrubbing sugars and salts, they will leave your lips feeling smooth and looking healthy and bright. They can be used one or twice per week.

174

Vegan alternative

If you don't like using beeswax in your recipes, then you can use soya wax, carnauba wax or candelilla wax instead in order to make your products vegan-friendly.

> ### FIX IT
>
> When making lip balms you may notice a small hole in the centre of your cooled product. Don't worry, as this is normal and does not affect the product at all. You can also gently warm the top with a hairdryer to re-melt the lip balm and fill in the hole.

175

Make an ultra-nourishing and protecting lip-care balm

Ingredients:
- *¼ teaspoon coconut oil*
- *¼ teaspoon olive oil*
- *¼ teaspoon mango butter*
- *½ teaspoon beeswax*
- *10 drops vitamin E oil*
- *1 drop lavender essential oil*
- *1 drop lemongrass essential oil*

Makes enough for one 4-ml lip balm tube

1. Measure out the coconut oil, olive oil, mango butter and beeswax into a small glass measuring jug and place in a saucepan containing a few centimetres of simmering water until the butter and beeswax have fully melted. Remove from the heat.

2. Add the vitamin E oil and two essential oils. Carefully pour the mixture to the very top of your lip balm container. If your hand isn't steady enough to pour melted lip balm into tiny tubes, use a disposable plastic pipette to fill them up.

3. Allow to harden and cool to room temperature

176

Different lip-care containers

Lip balm tubes: These cylindrical plastic containers feature a screwing bottom that pushes up the lip balm. Each small container can hold approximately 4 ml of lip product. You can find lip balm tubes in clear, white, black and a few other colours. They also come in round and oval shapes. If you wish to make larger amounts of lip balm, then a jumbo-sized lip balm tube is available that can hold 15 ml of lip balm. A few companies even sell environmentally friendly, paper, push-up lip balm tubes.

Lip balm jars and pots: These are small glass or plastic containers, which can hold 7–15 ml of lip product. They have either plastic or metal lids.

Lip balm tins: These small metal tins are available in either a round or rectangular shape. The round containers can be bought with either a rolled-edge cover or a twist top. The rectangular tins have a convenient slide top.

Recipes:
Three of the Best Lip-Care Treatments

177

Peppermint Cocoa Lip Balm

Makes enough for three 15-ml containers

This is a wonderful tingly treat for your lips, helping to repair and nourish them.

1 tablespoon, plus two teaspoons sweet almond oil
1 heaping tablespoon grated cocoa butter
1 tablespoon grated beeswax
10 drops vitamin E oil
15 drops peppermint essential oil

1. Measure out the sweet almond oil, cocoa butter and beeswax into a small glass measuring jug and place in a saucepan containing a few centimetres of simmering water until the butter and beeswax have fully melted. Remove from the heat.
2. Add the vitamin E and peppermint essential oils. Carefully pour the mixture to the very top of your containers. Allow to harden and cool to room temperature.

To use: Apply the desired amount to lips.

178

Fruity Lip Scrub

Makes approx. 30 ml

This is a sweet, strawberry-scented scrub to polish and protect your lips.

½ teaspoon beeswax
2 teaspoons sweet almond oil
1 teaspoon mango butter
1¼ teaspoons finely ground freeze-dried strawberries
2 teaspoons caster sugar
40 drops vitamin E oil

1. Measure out the beeswax, sweet almond oil and mango butter into a small glass measuring jug and place in a saucepan containing a few centimetres of simmering water until the beeswax and butter have fully melted.
2. Remove from the heat and quickly stir in the ground strawberries, sugar and vitamin E oil. When the mixture starts to harden, simply spoon it into a container with a lid to continue cooling to room temperature.

To use: Massage a small amount of the scrub over moistened lips to exfoliate and moisturise. Tissue or rinse off.

179

Terrific Tinted Lip Gloss

Makes approx. 85 ml of lip balm that can be divided between various-sized containers (round metal or glass containers work best for this recipe)

This rich and nourishing lip balm has a tint of colour.

1 tablespoon sweet almond oil
1 tablespoon coconut oil
1 tablespoon jojoba oil
1 tablespoon cocoa butter
1 heaping tablespoon grated beeswax
1 teaspoon vitamin E oil
1½ teaspoons powdered herb (choose from: powdered beetroot, powdered alkanet root, powdered cocoa or powdered hibiscus). Add more for a more intense colour.

1. Measure out the sweet almond oil, coconut oil, jojoba oil, cocoa butter and beeswax into a small glass measuring jug and place in a saucepan containing a few centimetres of simmering water until the butter and beeswax have fully melted. Remove from the heat.
2. Add the vitamin E oil and powdered herb. Stir to combine. Carefully and quickly pour the mixture into your lip balm containers. Allow to harden and cool to room temperature.

To use: Apply the desired amount to lips.

▼ Peppermint Cocoa Lip Balm, Fruity Lip Scrub and Terrific Tinted Lip Gloss

5 Conditioning Body Care

Conventional body-care products can be overloaded with harsh surfactants, potentially risky parabens, synthetic fragrances and other dubious ingredients. In this chapter, we will look at natural body cleansing, exfoliating and moisturising formulas that contain mild plant-based purifiers, sweet-smelling essential oils and nourishing plant and nut oils to nurture and protect your skin.

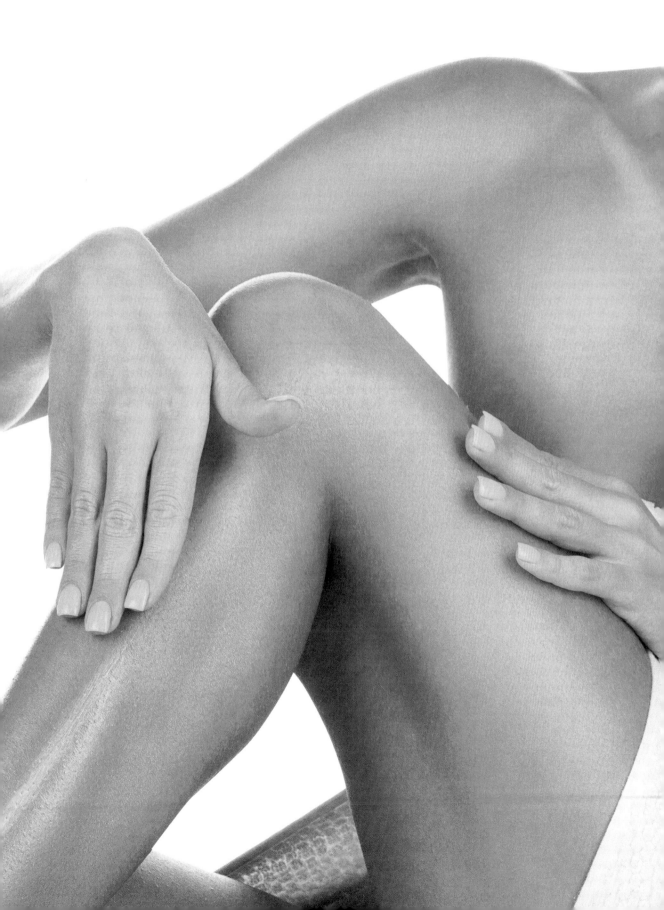

Body Cleansers

Infused with natural botanicals and essential oils, liquid body cleansers are a wonderful way to cleanse your skin gently and thoroughly, while also leaving it delicately scented with the soft aromas of natural essential oils.

180

Make a basic soap-based body cleanser

A basic soap-based cleansing recipe suitable for all skin types.

Ingredients:
- *3 tablespoons boiling water*
- *2 teaspoons plain table salt*
- *60 ml liquid Castile soap*

Makes approx. 90 ml

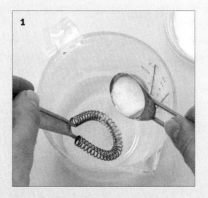

1. Add the boiling water to a small glass measuring jug. Add the salt and stir to dissolve. Set aside.

2. Add the liquid Castile soap to a sanitised bottle with a lid or pump.

3. Place a funnel in the bottle and add 2 tablespoons of the salt solution to the liquid Castile soap. Discard the remaining salt solution.

4. Place the lid on the bottle, and shake well to thicken. Store in the fridge and use within 2 weeks.

To use: Shake the bottle well and apply a small amount to wet skin, massage in with a flannel and rinse off with warm water.

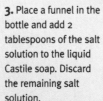

TRY IT

- Quickly massage a small amount of body oil over your moist skin after a shower to seal in moisture.

- Use natural body cleanser as a shave cream for perfect silky legs.

181

Getting it right for your skin type

Sensitive skin may benefit from the gentleness of body washes. Normal and oily skin may enjoy using bar soap.

182

Why cleanse natural?

It is important to remember that the purpose of body cleansing is to gently remove dirt, sweat, body odour and excess sebum and oil. You do not need to use harsh cleansers to get the job done. In fact, many traditional synthetic-laden cleansing products are harsh and can strip the skin of its natural oils, which in turn can leave your skin feeling parched and sensitive.

Gone are the days of using harsh soap made of animal lard and ash to cleanse your skin. Thankfully, there is a plethora of natural cleansing options available to us. From plant-based body washes to essential-oil-kissed bars of beautiful soap, whether you have sensitive skin or normal skin, there is a cleansing product that is just right for you.

183

Surfactants hold the secret

The magic part of any cleansing product is the surfactant ingredients. A surfactant cleanses your skin by dissolving the dirt and grime and allowing the water to wash it away from the skin. Essential oils and plant extracts are often added to cleansers to impart a nice smell or for a therapeutic benefit to the skin. Tea tree oil is commonly added into washes designed for blemish-prone skin. When purchasing natural body cleansers at a shop, look for the ingredient decyl glucoside, which is a mild surfactant made from plants and is biodegradable.

184

There's more to life than foam

Many natural cleansers will not foam as much as most synthetic-based cleansers. The surfactants used in the natural versions still cleanse just as well.

Recipes:
Four of the Best Body Cleansers

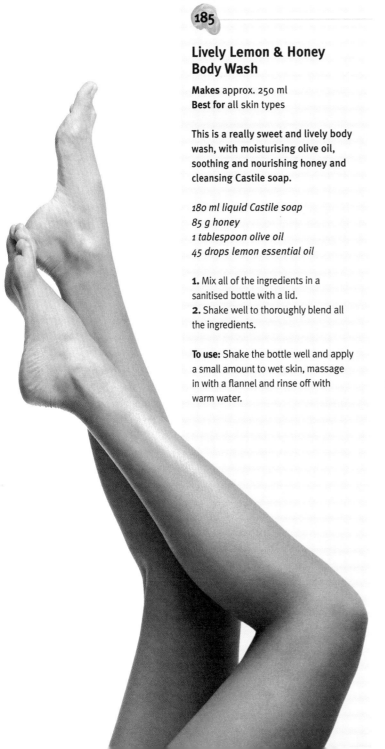

185

Lively Lemon & Honey Body Wash

Makes approx. 250 ml
Best for all skin types

This is a really sweet and lively body wash, with moisturising olive oil, soothing and nourishing honey and cleansing Castile soap.

180 ml liquid Castile soap
85 g honey
1 tablespoon olive oil
45 drops lemon essential oil

1. Mix all of the ingredients in a sanitised bottle with a lid.
2. Shake well to thoroughly blend all the ingredients.

To use: Shake the bottle well and apply a small amount to wet skin, massage in with a flannel and rinse off with warm water.

186

Patchouli & Vetiver Body Wash

Makes approx. 250 ml
Best for all skin types

Treat yourself with this woodsy and earthy scented body wash with jojoba oil, vegetable glycerin and Castile soap.

120 ml liquid Castile soap
60 ml vegetable glycerin
60 ml jojoba oil
10 drops vetiver essential oil
10 drops patchouli essential oil

1. Mix all of the ingredients in a sanitised bottle with a lid.
2. Shake well to thoroughly blend all the ingredients.

To use: Shake the bottle well and apply a small amount to wet skin, massage in with a flannel and rinse off with warm water.

187

Invigorating Peppermint & Rosemary Body Wash

Makes approx. 250 ml
Best for normal skin

This invigorating and cooling body wash is perfect for those hot summer days. The peppermint essential oil leaves a lingering cool feeling on the skin.

240 ml liquid Castile soap
15 drops peppermint essential oil
20 drops rosemary essential oil

1. Mix all of the ingredients in a sanitised bottle with a lid.
2. Shake well to thoroughly blend all the ingredients.

To use: Shake the bottle well and apply a small amount to wet skin, massage in with a flannel and rinse off with warm water. Avoid applying to sensitive areas of the body.

188

Sensitive-Skin Cleansing Wash

Makes approx. 250 ml
Best for sensitive skin

This soap-free and unscented recipe is fantastic for those with sensitive skin.

160 ml vegetable glycerin
60 ml castor oil
2 tablespoons sesame oil

1. Mix all of the ingredients in a sanitised bottle with a lid.
2. Shake well to thoroughly blend all the ingredients.

To use: Shake the bottle well and apply a small amount to wet skin, massage in with a flannel and rinse off with warm water.

189

Blemish-Blasting Body Wash

Makes approx. 200 ml
Best for normal and oily skin

A wonderful and purifying wash for those with blemishes. Contains the antibacterial power of thyme essential oil and tea tree essential oil.

3 tablespoons boiling water
2 teaspoons plain table salt
120 ml liquid Castile soap
2 tablespoons vegetable glycerin
60 drops tea tree essential oil
40 drops thyme essential oil

1. Add the boiling water to a small glass measuring jug.
2. Add the salt and stir to dissolve. Set aside.
3. Add the liquid Castile soap to a sanitised bottle with a lid or pump.
4. Place a funnel in the bottle and add 2 tablespoons of the salt solution to the liquid Castile soap. Discard the remaining salt solution.
5. Add the vegetable glycerin and the essential oils.
6. Place the lid on the bottle, and shake well to thicken. Store in the fridge and use within 2 weeks.

To use: Shake the bottle well and apply a small amount to wet skin, massage in with a flannel and rinse off with warm water.

Body Scrubs

Body scrubs are special treatments that exfoliate the skin, help increase blood flow, tighten the pores and promote a healthy, gorgeous glow. Body scrubs should be used in moderation (once to twice per week). They are spread over damp skin in soft, circular motions and rinsed away with warm water. At no time should you use them on broken or damaged skin.

190

Stay sweet

If you have sensitive skin, stick to sugar scrubs, as salt scrubs may irritate already sensitive skin.

192

Make a customised body scrub

You can make your own customised body scrubs with endless possibilities. To mix your own scrub, blend together your choice of one ingredient from each column below in a small bowl and then keep the scrub in a jar with a lid.

TRY IT

Replace the oil portion of the Basic Salt or Sugar Body Scrub (right) with mashed avocado or banana as a fun and nourishing alternative. Be sure to keep in the fridge and use within three days.

FIX IT

• If your body scrub is too thick, add a little more oil. If it is too thin, add more salt or sugar.

191

Make and use a basic salt or sugar body scrub

Ingredients:
• *2 parts fine-grain sugar or salt*
• *1 part oil (choose from coconut, olive, jojoba or sweet almond, etc.)*

1. Mix both ingredients together in a small bowl.
2. Apply to damp skin in soft, circular motions to exfoliate. Rinse away with warm water.

Caution: The oil in this recipe may make your shower or bath surface very slippery, so care needs to be taken not to slip or fall.

Add 200 g from this list	Add 100 g from this list	Add 160 ml from this list	Add up to 1 tablespoon from this list (optional)	Add up to ½ teaspoon from this list
• Brown sugar • Light brown sugar • Caster sugar • Dead Sea salt • Sea salt	• Coarse sugar • Demerara sugar • Muscovado sugar • Epsom salts • Ground oats • Maize flour • Apricot kernel meal • Bicarbonate of soda	• Jojoba oil • Coconut oil • Sweet almond oil • Grapeseed oil • Kukui nut oil • Olive oil • Pumpkin seed oil • Mashed banana (use the scrub the same day) • Mashed avocado (use the scrub the same day) • Yoghurt (use the scrub the same day)	• Ground cinnamon • Cocoa powder • Lemon zest • Orange zest • Lavender buds • Honey • Ground rice powder • Coconut flakes • Ground freeze-dried fruit • Sesame seeds • Honey • Ground herbs or flowers • Espresso powder	• Essential oil (single or blend) • Vanilla extract • Natural flavouring • Natural food colouring

Recipes:
Three of the Best Body Scrubs

193

Key Lime Pie Body Scrub

Makes approx. 340 g
Best for approx. 350 ml

You'll just love this sugary-sweet treat that smells like freshly baked home-made Key lime pie.

80 ml coconut oil
2 tablespoons sweet almond oil
2 tablespoons vegetable glycerin
110 g packed dark brown sugar
70 g caster sugar
30 drops lemon essential oil
30 drops lime essential oil

1. Combine the coconut oil, sweet almond oil and vegetable glycerin in a small bowl.
2. Add the brown sugar and white sugar, and mix well.
3. Stir in the essential oils.
4. Transfer to a jar with lid.

To use: Massage a small amount over damp skin and rinse away with warm water.

194

Heavenly Herbal Purifying Scrub

Makes 1 full-body scrub application
Best for all skin types

Reveal a glowing and radiant appearance with this mellow herb- and flower-based scrub that softly sloughs away dull skin.

60 g rhassoul clay
25 g rolled oats
4 tablespoons dried lavender buds
4 tablespoons dried rose petals
40 g maize flour
40 g ground almonds
240 ml plain yoghurt
170 g honey
60 drops lavender essential oil
20 drops Moroccan chamomile
  essential oil

1. Place the rhassoul clay, rolled oats, lavender buds and rose petals in a food-processor and pulse to grind into a fine powder.
2. Transfer to a mixing bowl, add the maize flour and ground almonds, and then stir to combine.
3. Stir in the yoghurt, honey and essential oils, and mix well to combine. (Add a small amount of warm water if the mixture is too thick.)

To use: Stand in the shower and moisten your skin with warm water. Turn off the water and massage the scrub over your entire body, from the neck down, using gentle, circular motions. Leave on for 5 minutes and rinse away with warm water.

195

Sweet Orange & Banana Sugar Scrub

Makes 1 full-body scrub application
Best for all skin types

This is the ultimate soothing sugar scrub for all skin types. It smells incredible as a result of the fresh banana and orange essential oil.

1 small ripe banana
50 g caster sugar
1 tablespoon vegetable glycerin
40 drops sweet orange essential oil

1. Mash the banana in a small bowl until chunky.
2. Stir in the sugar, vegetable glycerin and orange essential oil.

To use: Apply the scrub to damp skin in soft, circular motions. Rinse away with warm water.

► Sweet Orange &
Banana Sugar Scrub

Body Moisturisers

Body moisturisers are put on after a body cleansing and exfoliating ritual.
These marvellous moisturisers protect and soothe the skin. Moisturising after
a bath helps the skin maintain its optimal moisture levels during the day, so
keeping it soft and supple.

196

Choosing the best body-moisturising product

Body lotions: These are light moisturising fluids that are easily absorbed into the skin. They can soften and lightly scent the skin if essential oils are also added to the recipe. Body lotions are usually made with lighter carrier oils and are thin enough to be used in a pump bottle. Good for all skin types.

Body creams: Much like lotions, although body creams are significantly creamier and denser. They are formulated with both carrier oils and butters (such as shea butter). Body cream adds a protective barrier to the skin's top layer, which helps to restore softness, especially to chapped and parched skin. A body cream is the perfect solution for protecting the skin from cold weather. They are usually too thick for a pump bottle and so are best stored in a jar.

Body butters: Even thicker and richer than body creams, body butters are highly concentrated formulas that are not as easily absorbed as lotions or creams. They are best suited to super-dry skin where a barrier cream is required. They are usually formulated with high amounts of natural butters such as shea, cocoa or mango butter.

Body balms: These are similar to a salve, where a herb-infused carrier oil is thickened with beeswax. Balms do not absorb as readily and are meant to protect the top layer of the skin by acting as a barrier against moisture loss.

Solid lotion bars: As the name implies, solid lotion bars are convenient bars of skin-softening ingredients, such as natural butters and carrier oils, which are thickened with beeswax and often scented with essential oils. You simply rub the lotion bar over your skin and then massage in. They come in fun shapes and sizes, and are portable if carried in a tin or container. You can use them on your lips, cuticles, feet and, of course, your body.

Body oils: These are liquid body moisturisers made from carrier oils and essential oils, which are perfect for massage. They are usually formulated along with therapeutic essential oils. Fast-absorbing and rejuvenating carrier oils include jojoba, sweet almond, grapeseed, sunflower, sesame and coconut oils.

197

Make a customised solid lotion bar

Solid lotion bars are a super-fun beauty treat to make. You will need a silicone mould for the best results. You can easily customise your own lotion bars using the instructions and ingredient choices given below.

Makes approx. 280 g of lotion bars

1. Begin by adding your choice of ingredients from Columns 1 and 2 to a glass measuring jug, placed in a simmering water bath until they have melted.

2. Add your choice of ingredients from Columns 3 and 4, and allow these to melt completely.

3. Remove from the heat and stir in your selection of essential oils from Column 5. (Adding essential oils at this stage is optional.)

5. Pop out the lotion bars from the mould and store in tin, metal or glass containers at room temperature.

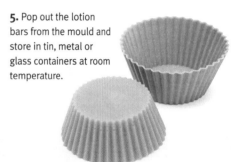

4. Carefully pour the hot mixture into the cavities of your silicone moulds. Allow to harden completely to room temperature.

Column 1	Column 2	Column 3	Column 4	Column 5
Choose 5 tablespoons from this list	**Use 3 tablespoons from this list**	**Choose 3 tablespoons from this list**	**Choose 5 tablespoons from this list**	**Use up to 120 drops of essential oils (optional)**
• Beeswax • Candelilla wax (use only 3 tablespoons of this wax) • Carnauba wax • Soya wax	• Cocoa butter	• Illipe butter • Kokum butter • Mango butter • Shea butter	• Almond oil • Avocado oil • Baobab oil • Coconut oil • Grapeseed oil • Hazelnut oil • Hemp seed oil • Herbal-infused carrier oil • Macadamia nut oil • Olive oil • Sesame oil • Sunflower oil	Suggested blends: **Insomnia blend:** 18 drops Roman chamomile, 30 drops lavender, 16 drops sweet orange, 12 drops ylang ylang **Euphoria blend:** 12 drops lemon, 24 drops lime, 12 drops peppermint, 25 drops petitgrain, 25 drops rosemary **Happy blend:** 25 drops lavender, 25 drops vanilla absolute, 30 drops rose absolute, 13 drops jasmine absolute

Recipes:
Five of the Best Body Moisturisers

198

Lavender & Sandalwood Body Lotion

Makes approx. 220 ml
Best for all skin types

A double dose of lavender makes this the perfect lotion to use before bed for sweet slumber.

200 ml lavender floral water (hydrosol)
2 teaspoons vegetable glycerin
2 tablespoons sweet almond oil
1 teaspoon stearic acid
1 teaspoon liquid lecithin
2 teaspoons emulsifying wax NF
1 teaspoon vitamin E oil
60 drops lavender essential oil
40 drops Australian sandalwood essential oil

1. Measure the lavender floral water and vegetable glycerin into a glass measuring jug and sit this in a saucepan containing a few centimetres of simmering water.
2. Measure the sweet almond oil, stearic acid, liquid lecithin and emulsifying wax into a glass measuring jug and sit it in a saucepan containing a few centimetres of simmering water.
3. When both mixtures have reached a temperate 71°C (160°F), remove them from the heat.
4. Carefully pour the oil mixture into a heatproof mixing bowl and begin mixing with a hand-held mixer set on medium speed.
5. Carefully add the floral water and glycerin, and continue to mix for 5 minutes.
6. Add the vitamin E and essential oils, and mix well.
7. Transfer the finished lotion to a sanitised bottle with a pump or lid. Store in the fridge and use within 15 days.

199

Velvet Moisture Body Butter

Makes approx. 200 ml
Best for all skin types

Made with shea butter and mango butter, this recipe feels like velvet on your skin. A very hydrating and protecting formula.

120 ml distilled water
1 tablespoon avocado oil
1 tablespoon coconut oil
1 tablespoon shea butter
1 tablespoon mango butter
2 teaspoons stearic acid
1 tablespoon emulsifying wax NF
1 teaspoon vitamin E oil

1. Measure the distilled water into a glass measuring jug and sit this in a saucepan containing a few centimetres of simmering water.
2. Measure the avocado oil, coconut oil, shea butter, mango butter, stearic acid and emulsifying wax into a glass measuring jug and sit it in a saucepan with a few centimetres of simmering water.
3. When both mixtures have reached a temperature of 71°C (160°F), remove them from the heat.
4. Carefully pour the oil mixture into a heatproof mixing bowl and begin mixing with a hand-held mixer set on medium speed.
5. Carefully add the distilled water and continue to mix for 5 minutes.
6. Add the vitamin E and mix well.
7. Transfer the finished cream to a sanitised glass jar. Store in the fridge and use within 15 days.

200

Skin-Soothing Calendula & Palmarosa Body Balm

Makes approx. 130 ml
Best for all skin types

Calendula-infused oil is gentle and soothing to the skin, and can assist with dry and parched skin, inflammation, blemishes and other skin complaints.

85 ml calendula-infused carrier oil
1 tablespoon, plus 1$\frac{1}{2}$ teaspoons beeswax
20 drops palmarosa essential oil
15 drops geranium essential oil

1. Add the calendula-infused carrier oil and beeswax to a glass measuring jug and sit this in a saucepan containing a few centimetres of simmering water.
2. Once the oil and wax have completely melted, remove from the heat.
3. Stir in the essential oils.
4. Transfer the body balm to a glass jar or metal tin.
5. Cover and allow to cool completely. Use within 6 months.

201

Nourishing Lavender Body Oil

Makes approx. 120 ml
Best for all skin types

Lightly scented with pure lavender essential oil, this is perfect as a moisturising massage oil!

2 tablespoons jojoba oil
2 tablespoons sweet almond oil
2 tablespoons sunflower oil
2 tablespoons avocado oil
1 teaspoon vitamin E oil
50 drops lavender essential oil

1. Mix all of the ingredients into a pump or lidded bottle. Massage a generous amount into the skin. Use within 6 months.

202

Ultra-Rich Carrot Seed & Helichrysum Body Cream

Makes approx. 225 ml
Best for all skin types

This is the ultimate body cream recipe. Fashioned with three luxurious oils, shea butter and both carrot seed and skin-rejuvenating Helichrysum essential oil, this body cream will transform dry parched skin into radiant supple skin.

160 ml rose floral water
1$\frac{1}{2}$ teaspoons vegetable glycerin
1 tablespoon olive oil
1 tablespoon argan oil
1 tablespoon pomegranate seed oil
1 tablespoon shea butter
2 teaspoons stearic acid
1 tablespoon emulsifying wax NF
1 teaspoon vitamin E oil
45 drops carrot seed essential oil
25 drops Helichrysum essential oil

1. Measure the rose floral water and vegetable glycerin into a glass measuring jug and sit this in a saucepan containing a few centimetres of simmering water.
2. Measure out the olive oil, argan oil, pomegranate seed oil, shea butter, stearic acid and emulsifying wax into a glass measuring jug and sit it in a saucepan containing a few centimetres of simmering water.
3. When both mixtures have reached a temperatue of 71°C (160°F), remove them from the heat.
4. Carefully pour the oil mixture into a heatproof mixing bowl and begin mixing with a hand-held mixer set on medium speed.
5. Carefully add the floral water and glycerin, and continue to mix for 5 minutes.
6. Add the vitamin E and essential oils, and mix well.
7. Transfer the cream to a sanitised glass jar. Store in the fridge and use within 15 days.

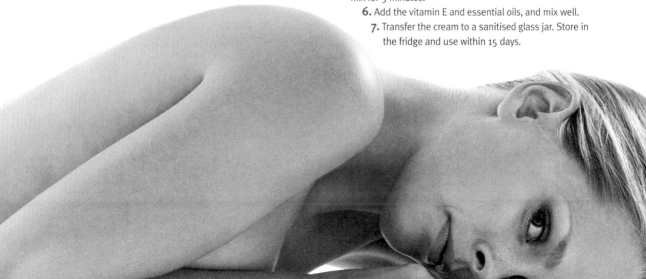

203

Using bath salts and soaks

These are typically made with salts such as Epsom salts, Dead Sea salts and Himalayan bath salts, along with other ingredients like powdered herbs, clays and essential oils. Soaking in a bath enhanced with a special bath salt can boost the immune system, energise the body and even relieve sore muscles. These also make wonderful gifts when packaged in pretty tubs.

204

Using bath oils

Bath oils are wonderful, especially if you have dry or sensitive skin. Comforting carrier oils, along with aromatic essential oils, will protect and richly moisturise the skin. Bath oils are highly concentrated and only a few teaspoons are recommended per bath. Always add bath oil to a bath after it has filled with warm water to prevent the essential oils evaporating quickly.

205

Using bath teas

As the name implies, bath teas are a blend of herbs and flowers, which are tied in sachets that infuse bath water with therapeutic magic to help soothe sore muscles, calm frazzled nerves, relieve dry itchy skin and help relieve congestion. The best way to use a bath tea is to add a cup of bath tea to a 10-cm by 15-cm cotton muslin bag with the drawstring tightly drawn. Add the bag to the bath while it fills with warm water. Allow the bath tea to infuse the water, while you bathe. Discard the spent herbs into your compost pile and rinse and dry the cotton muslin bag to reuse.

Bath Treatments

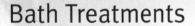

Pamper your skin, soothe your soul and indulge your senses with nutrient-rich bath recipes that softly cleanse, revive and calm your body. Choose from bath salts and soaks, fizzy bath bombs, bath teas and bath milks.

206

Using milk baths

These are indulgent beautifying baths that infuse the water with moisturising milk, which contains lactic acid to help beautify the skin and exfoliate dead skin cells. You simply fill your bath with warm water and add the milk bath, swirling to mix, and then relax for 20 to 30 minutes. Cow's milk and goat's milk are the most common ingredients, but you can also use non-dairy options such as almond, soya and hemp milk. You may use fresh dairy if you're using the milk bath the same day, or use powdered dry milk for blends that do not require refrigeration and for gift-giving. Substitute 3 tablespoons of powdered milk in place of every 240 ml of fresh milk and 70 g of powdered coffee creamer in place of 240 ml of cream.

Caution: Surfaces may become slippery when you are using milk baths or bath oils, so care should be taken when stepping in and out of the bath.

207

Make your own fizzy bath bombs

Make bath-time super fun! The secret ingredients in bath bombs are bicarbonate of soda and citric acid which, when mixed with water, start to fizz and bubble. Bath bombs can be customised with moisturising carrier oils, powdered herbs and flowers, essential oils and even natural food colours. To make bath bombs, you will need a selection of dry and wet ingredients (simply choose a combination of ingredients from the recipe builder below), plus suitable moulds such as silicone moulds, sweet moulds, an ice-cube tray, a muffin tin or soap moulds to shape your bombs.

Column 1	Column 2	Column 3	Column 4	Column 5	Column 6	Column 7	Column 8
Use 130 g of:	Use 65 g of:	Use 135 g of one of the following:	Use up to 1 tablespoon of one of the following:	Use 2 teaspoons of one of the following:	(Optional) Use 2 drops of:	(Optional) Use 60 drops of:	Use up to 1 tablespoon of the following to moisten the dry ingredients:
• Bicarb of soda	• Citric acid	• Fine Epsom salts • Fine Dead Sea salts • Fine Himalayan bath salts • Fine grey sea salt	• Powdered rose petals • Lavender buds • Powdered herbs • Powdered fruit peels	• Sweet almond oil • Olive oil • Grapeseed oil • Sesame oil • Sunflower oil • Jojoba oil	• Natural food colour	• Essential oils (blend your favourites together to total 60 drops)	• Witch hazel • Floral water • Vanilla extract • Vodka • Distilled water • Tinctures or herbal extracts

1. Combine your choice of dry ingredients in a large mixing bowl. Slowly drizzle in your choice of oil, essential oils and food colouring (if using), and mix well.

2. Put the wet ingredients – the citric acid and your choice of moistening ingredient – in a spray bottle and slowly spritz these over the dry ingredients, while constantly stirring to stop the ingredients fizzing.

3. Once all of the ingredients are moist enough to clump together, firmly press the mixture into the cavities of the moulds.
4. Allow the bombs to dry for 3–4 hours before unmoulding. Place them in an air-tight container to stop them being activated by the humidity in the air. Use within 4 months.

To use: Fill the bath with warm water and get in. Add 1–2 bath bombs (depending on the size) and enjoy.

Recipes:
Eight of the Best Bath Treatments

208

On Cloud Nine Bath Tea

Makes enough for 1 bath

Slip into sweet serenity with this tension-erasing recipe.

4 tablespoons jasmine flowers
4 tablespoons crushed rose
  petals
4 tablespoons chamomile flowers
1 tablespoon finely minced
  lemongrass
1 teaspoon lemon zest
1 teaspoon orange zest
2 teaspoons finely grated fresh
  ginger

1. Mix the herbs together and place them in a large, cotton muslin bag.

To use: Add the bag to the bath while it fills with warm water. Allow the bath tea to infuse the water, while you relax and bathe.

209

Skin-Soothing Bath Tea

Makes enough for 1 bath

Calm and comfort sensitive skin with soothing oats and calming calendula.

3 tablespoons ground oats
2 tablespoons calendula flowers
2 tablespoons marshmallow root
2 tablespoons chickweed
2 tablespoons dried lavender
2 tablespoons dried chamomile
  flowers

1. Mix the herbs together and place them in a large, cotton muslin bag.

To use: Add the bag to the bath while it fills with warm water. Allow the bath tea to infuse the water, while you relax and bathe.

210

Citrus-Splash Bath Oil

Makes approx. 60 ml

A refreshing and invigorating bath oil to awaken your senses and nourish your skin.

1 tablespoon grapeseed oil
1 tablespoon sunflower oil
1 tablespoon apricot kernel oil
1 teaspoon liquid lecithin
1 teaspoon castor oil
1 teaspoon vitamin E oil
30 drops sweet orange
  essential oil
30 drops lemon essential oil
30 drops lime essential oil
15 drops bergamot essential oil

1. Mix all of the ingredients in a glass bottle and shake well.

To use: Fill the bath with warm water. Add 2–3 teaspoons of bath oil and swirl to mix.

Note: You can substitute 90 drops of lavender essential oil in place of the citrus oils for a relaxing blend.

211

Hot Chocolate Milk Bath

Makes enough for 1 bath
Best for all skin types

Experience the delicious and decadent aroma of chocolate minus the calories! A fun treat for a kid's bath.

240 ml whipping cream
240 ml whole milk
60 g cocoa powder
1 tablespoon vanilla extract

1. Mix all of the ingredients in a small mixing bowl.

To use: Fill the bath with warm water, pour the milk bath into the water and swirl with your hand to blend. Soak and relax for 30 minutes. Rinse well after draining the bath water.

212

Beautifying Buttermilk Bath

Makes enough for 1 bath

Buttermilk will help you maintain soft and supple skin, and this recipe will leave you smelling sweet as a rose.

120 g powdered buttermilk
240 ml whole milk
85 g honey
2 teaspoons vitamin E oil
5 drops neroli essential oil
5 drops Bulgarian rose essential oil

1. Mix all of the ingredients in a small mixing bowl.

To use: Fill the bath with warm water, pour the milk bath into the water and swirl with your hand to blend. Soak and relax for 30 minutes. Rinse well after draining the bath water.

213

Fresh Queen Bee Milk Bath

Makes enough for 1 milk bath
Best for all skin types

Your skin will be soft and moisturised after a dip in this creamy milk bath.

240 ml whole milk
120 ml whipping cream
85 g honey
25 g finely ground oats
1 tablespoon vanilla extract

1. Mix all of the ingredients in a small mixing bowl.

To use: Fill the bath with warm water, pour the milk bath into the water and swirl with your hand to blend. Soak and relax for 30 minutes. Rinse well after draining the bath water.

214

Relaxing Lavender Bath Soak

Makes enough for 1 bath

Sweet and sound slumber is just a soak away with this lavender-infused recipe.

275 g medium-grade Dead Sea salts
30 g powdered dry milk
1 tablespoon vegetable glycerin
20 drops lavender essential oil

1. Add the Dead Sea salts and powdered dry milk to a small glass bowl, and stir well to combine.
2. Drizzle in the vegetable glycerin and stir well to combine.
3. Add the lavender essential oil and mix well.

To use: Add to a warm bath and soak for 30 minutes.

215

Super-Detox Bath Salt

Makes enough for 1 bath

A great recipe to soothe sore muscles and deodorise your skin.

135 g small- to medium-grade Himalayan bath salts
135 g Epsom salts
1 tablespoon bentonite clay
1 tablespoon kelp powder
1 teaspoon Matcha powder (finely ground green tea leaves)
1 teaspoon vitamin E oil
10 drops rosemary essential oil
10 drops lavender essential oils

1. Add the Himalayan and Epsom salts, bentonite clay, kelp powder and Matcha powder to a small glass bowl and stir well to combine.
2. Drizzle in the vitamin E oil and stir again to combine.
3. Add the essential oils and mix well.

To use: Add to a warm bath and soak for 30 minutes.

Caution: Surfaces may become slippery when you are using these recipes, so care should be taken when stepping in and out of the bath.

Underarm Deodorants

Natural deodorant works to counteract the growth of bacteria, eliminate harsh body odour and balance your sensitive underarm area without clogging your pores. Many traditional deodorants and antiperspirants contain aluminium, as well as other harsh and unnatural ingredients that may potentially do more harm than good. Making your own deodorant allows you to customise it with essential oils to create your very own signature blend.

216

Common ingredients in natural deodorants

Alcohol: Antibacterial liquid and carrier for essential oils and extracts in liquid deodorant recipes.

Aloe vera gel: Soothing liquid base for liquid deodorant recipes.

Arrowroot powder: Helps to absorb moisture and wetness. It comes from the plant *Maranta arundinacea*. Used in solid and powdered deodorant formulas.

Bicarbonate of soda: Helps to absorb perspiration and neutralise unpleasant odour molecules.

Cocoa butter: Thickens solid deodorant recipes and moisturises the skin.

Coconut oil: Thickens solid deodorant recipes and moisturises the skin.

Cornflour: Helps absorb moisture and wetness. Used in solid and powdered deodorant formulas.

Essential oils: Add a pleasant smell to the underarms, as well as provide antibacterial protection against odour-causing bacteria. The underarms are very sensitive areas and essential oils should be added in 1.5% dilution ratio or less to deodorant blends.

Kaolin clay: Helps absorb moisture from the skin. Used in solid and powdered deodorant formulas.

Shea butter: Thickens solid deodorant recipes and moisturises the skin.

Vegetable glycerin: Soothing liquid base for binding other ingredients and moisturising the skin.

Vitamin E oil: Skin conditioner and antioxidant to help preserve the other oils in your deodorant recipes.

Witch hazel extract: Antibacterial liquid and carrier for essential oils and extracts in liquid deodorant recipes.

217

Salt stones

Natural mineral salt stones are often used as natural deodorants. They are made from mineral salt deposits and are polished until they have smooth, rounded tops, which are then moistened with water and rolled onto your underarms. They are very effective and one 85-g stone can last up to a whole year.

218

Reapply

Natural deodorants are not antiperspirants, and may need to be reapplied during the day to help control body odour.

219

Give it time

It may take a few days to adjust to a natural deodorant if you are switching from conventional shop-bought brands. Just remember that natural deodorants feel and work differently than what you may be used to. Give your body and mind a few days to adjust and you will be happy that you switched to a healthier alternative.

Recipes:
Four of the Best Natural Deodorants

220

Unscented Natural Cream Deodorant

Makes approx. 85 ml
Best for all skin types

This recipe may be added to an empty, twist-up deodorant tube before it cools for easy application.

1 tablespoon shea butter
1 tablespoon coconut oil
1 tablespoon cocoa butter
1 tablespoon, plus 1 teaspoon bicarbonate of soda
1 tablespoon arrowroot powder
½ teaspoon kaolin clay
½ teaspoon vitamin E oil

1. Add the shea butter, coconut oil and cocoa butter to a small glass measuring jug and place this in a saucepan containing a few centimetres of simmering water until melted.
2. Remove from the heat and stir in the bicarbonate of soda, arrowroot powder, kaolin clay and vitamin E oil until smooth.
3. Pour into an empty, twist-up deodorant tube or small jar.

To use: Apply a small amount into each underarm. May be reapplied after heavy sweating or exercise.

221

Rose & Sandalwood Powder Deodorant

Makes approx. 60 g
Best for all skin types

A simple dusting under your arms will help you smell sweet the entire day. Helps absorb moisture too.

1 tablespoon cornflour
1 tablespoon arrowroot powder
1 teaspoon kaolin clay
1 teaspoon bicarbonate of soda
10 drops sandalwood essential oil
5 drops rose essential oil or rose absolute oil

1. Sift the cornflour, arrowroot powder, kaolin clay and bicarbonate of soda through a sieve into a bowl.
2. Drizzle in the essential oils, while stirring with a whisk.
3. Sift the ingredients once more to break up any lumps that may have formed while adding the essential oils.
4. For the best results, store the powder in a small sugar or salt shaker.

To use: Shake a small amount of powder on each underarm. The deodorant may be applied after you have used a body moisturiser to help keep it in place.

222

Cooling Peppermint Summer Deodorant Spray

Makes approx. 65 ml
Best for all skin types

This is the perfect deodorant for hot weather!

1 tablespoon witch hazel extract
1 tablespoon Aloe vera gel
2 tablespoons peppermint floral water
1 teaspoon vegetable glycerin
5 drops peppermint essential oil

1. Mix all the ingredients in a small spray bottle.

To use: Shake well, and spray on your underarms as needed throughout the day.

223

Lemon & Sage Deodorant Spray

Makes approx. 60 ml
Best for all skin types

Mist on your underarms as needed throughout the day.

3 tablespoons witch hazel extract
2 teaspoons sage tincture (extract)
1 teaspoon vegetable glycerin
10 drops tea tree essential oil
15 drops lemon essential oil

1. Mix all of the ingredients in a small spray bottle.

To use: Shake well, and spray on your underarms as needed throughout the day.

Manicures and Pedicures

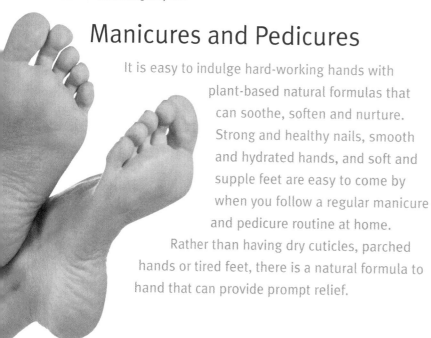

It is easy to indulge hard-working hands with plant-based natural formulas that can soothe, soften and nurture. Strong and healthy nails, smooth and hydrated hands, and soft and supple feet are easy to come by when you follow a regular manicure and pedicure routine at home. Rather than having dry cuticles, parched hands or tired feet, there is a natural formula to hand that can provide prompt relief.

225

Be kind to your cuticles

If you suffer from dry or cracked cuticles, keep them soft and supple by rubbing in a little olive oil several times a day.

224

Performing a once-a-week manicure and pedicure routine

Try to treat yourself to a manicure and pedicure once a week, following these simple instructions:

1. Remove any nail polish, as necessary.
2. Fill a large bowl with warm water or foot soak (see page 125). Soak your feet for 15–20 minutes to cleanse and soften skin. Rinse with warm water if using a foot soak.
3. Fill a small bowl with some warm water. Add ½ teaspoon of hand soap and ½ teaspoon of vegetable glycerin, and swirl to mix. Soak your fingertips in the warm water for 5 minutes to soften the cuticles and remove dirt and surface stains. Rinse well with warm water.
4. Apply a sugar- or salt-based scrub and massage over your hands and fingers, with a gentle pressure, to exfoliate and moisturise. Rinse well with warm water. Repeat this step on your feet. Scrubs work well to remove stains, dirt and dead skin.
4. Apply a small amount of cuticle cream or carrier oil (such as olive or sweet almond oil) to your cuticles and nail beds. Gently push them back with a manicure stick. Never cut your cuticles with nail clippers or scissors.
5. Trim your nails with clippers if necessary to the desired length.
6. Use a nail buffer to smooth out any ridges on the surface of the nails.
7. Use a nail file to gently file the tips of your nails to the desired length.
8. Massage your hands and feet with a generous amount of thick hand cream. Do not rinse off.

◀ Use a nail file, clippers and a manicure stick to perform the perfect manicure.

226

Multi-use moisturisers

Use hand lotions and creams over your entire body to keep it hydrated and supple.

227

Treating your toes

Your feet work very hard and can get worn out in a hurry. This may be because you've walked too long, cramped them in tight-fitting shoes, stood on them all day or they've got too hot – they deserve a refreshing break to relax and recover. The recipes for the customised hand salve, hand lotion and hand cream over the page are also ideal for using on your feet.

228

Make an argan oil and pomegranate fortifying cuticle and nail cream

This is a very rich and creamy treat for the nails and cuticles that will condition and protect.

Ingredients:
- *1 tablespoon shea butter*
- *1 tablespoon lanolin oil*
- *1 heaping tablespoon beeswax*
- *1 tablespoon pomegranate seed oil*
- *1 tablespoon argan oil*
- *¼ teaspoon vitamin E oil*
- *3 drops rose absolute*

Makes approx. 40 ml

1. Measure the shea butter, lanolin oil and beeswax into a glass measuring jug and sit this in a saucepan containing a few centimetres of simmering water, stirring occasionally, until the ingredients have melted.

2. Remove from the heat and, while the measuring jug is still in the hot water, stir in the pomegranate seed oil, argan oil, vitamin E oil and rose absolute.

3. Remove the measuring jug from the water and use a towel to remove any moisture from the outside.

4. Carefully pour the mixture into a small, glass jar or tin container.

To use: Apply a small amount to the base of each nail and massage into the cuticle and nail.

229

Choosing products to keep hands in tip-top shape

You can use a range of home-made beauty products to treat your hands and nails, including:

Hand scrubs: These are usually either sugar- or salt-based and combined with conditioning carrier oils to slough away dead skin cells, dirt, stains and odours, so revealing silky, supple and smooth hands. If you have sensitive hands, a sugar scrub is the best choice, because salt-based scrubs can sting or irritate delicate skin. Hand scrubs may be used several times per week.

Hand sanitisers: The best hand sanitiser is the formula that contains at least 65% ethanol. Several of the other commercial hand sanitisers available contain dubious chemicals that may pose health risks. A bit of vegetable glycerin and some added essential oil, along with the ethanol, will make a wonderfully scented and moisturising formula.

Nail and cuticle oils: These products are used to revitalise dry, easily broken and cracked nails. They are composed of luxurious carrier oils and essential oils. Nail and cuticle oils care for the cuticle and promote strong, healthy nails. You simply massage one drop of oil into each nail before going to bed. These treatments can also be applied to soften the cuticles during a manicure.

Hand lotions, creams and salves: These formulas work to relieve hands from dryness and to protect the hands. Salves are often made with herb-infused oils. Hand lotions, creams and salves should be applied after each hand washing and before bed for the best results.

230

Just add oil

You can add up to 50 drops of essential oil to each of the recipes over the page. Here are a few blends to try out:
- **Dry skin blend:** 30 drops lavender essential oil, 10 drops Palmarosa essential oil, 5 drops German chamomile essential oil and 5 drops of carrot seed essential oil.
- **Healing hands blend:** 20 drops lavender essential oil, 10 drops tea tree essential oil, 10 drops Helichrysum essential oil and 5 drops German chamomile essential oil.
- **Heavenly smelling hands blend:** 10 drops rose geranium essential oil, 5 drops rose absolute, 5 drops jasmine absolute, 5 drops ylang ylang essential oil, 5 drops vetiver essential oil and 5 drops Australian sandalwood essential oil.

231

Make a customised hand salve

You will need to prepare a herbal-infused oil first (see pages 26–27), as this makes up one of the ingredients in the hand salve, before selecting your choice of other ingredients from the chart below.
Makes approx. 325 ml

1. Measure the herbal-infused carrier oil(s) and your chosen wax into a glass measuring jug and sit this in a saucepan containing a few centimetres of simmering water, stirring occasionally, until melted.
2. Remove from the heat.
3. Stir in the vitamin E oil and essential oils.
4. Carefully pour the mixture into a small, glass jar or tin container.

To use: Apply the salve to your hands and massage in as needed.

Herbal-Infused Carrier Oils Use 120 ml of carrier oil (choose a combination of as many different oils as you wish):	Wax Use 2 heaping tablespoons of one of the following:	Vitamin E oil Use ¼ teaspoon of:	Essential oils Choose a total of 10 drops of:
• Burdock (*Arctium lappa*) infused carrier oil • Calendula (*Calendula officinalis*) infused carrier oil • Comfrey (*Symphytum officinale*) leaf/root infused carrier oil • Marshmallow (*Althaea officinalis*) root infused carrier oil • Mullein (*Verbascum*) leaf infused carrier oil • Nettle (*Urtica dioica*) leaf infused carrier oil • Plantain (*Plantago major*) infused carrier oil • Yarrow (*Achillea millefolium*) leaf/flower infused carrier oil	• Beeswax (grated) • Carnauba wax (flaked)	• Vitamin E oil	• Australian sandalwood • Cistus • Eucalyptus • Frankincense • Geranium • German chamomile • Helichrysum • Lavender • Moroccan chamomile • Myrrh • Patchouli • Peppermint • Rosemary • Tea tree • Other favourite essential oils

232

Make a customised hand cream

Create your very own ultra-rich hand cream to provide intense hydration and protection for even the driest parched

skin. This formula-builder gives you a very thick product, which will need to be stored in a pot or jar and then scooped out.
Makes approx. 120 ml

1. Measure your chosen ingredients from Columns A, B, C and D into a glass measuring jug and sit this in a saucepan containing a few centimetres of simmering water, stirring occasionally, until everything has melted and the temperature reaches about 71°C (160°F).
2. Measure your choice of ingredients from Column E into another glass measuring jug, sit this in a saucepan of simmering water and heat to a temperature of about 71°C (160°F).

3. Carefully remove both measuring jugs from the simmering water.
4. Pour the oil and wax mixture carefully into a heatproof mixing bowl and begin mixing with a hand-held mixer set on medium speed.
5. Carefully add the water mixture and continue to mix for a further 5 minutes.
6. Once the mixture has cooled to under 38°C (100°F), mix in any optional ingredients from Column F.
7. Pour your lotion into a sanitised container and allow to cool completely and thicken. Store the cream in the fridge and use within 2 weeks.

Make a customised hand lotion

Create your very own lightweight, fast-absorbing hand lotion that will sink quickly into your skin to nourish, smooth and moisturise. This formula is light and fluffy enough to store in a bottle with a pump. Simply select the ingredients for the hand lotion from those listed in the chart below.

Makes approx. 120 ml

1. Measure your chosen ingredients from Columns A, B, C and D into a glass measuring jug and sit this in a saucepan containing a few centimetres of simmering water, stirring occasionally, until everything has melted and the temperature reaches about 71°C (160°F).
2. Measure your choice of ingredients from Column E into another glass measuring jug, sit this in a saucepan of simmering water and heat to a temperature of about 71°C (160°F).
3. Carefully remove both measuring jugs from the simmering water.
4. Pour the oil and wax mixture carefully into a heatproof mixing bowl and mix with a hand-held mixer set on medium speed.
5. Carefully add the water mixture and mix for a further 5 minutes.
6. Once the mixture has cooled to under 38°C (100°F), mix in any optional ingredients from Column F.
7. Pour your lotion into a sanitised container and allow to cool completely and thicken. Store the lotion in the fridge and use within 2 weeks.

Column A	Column B	Column C	Column D	Column E	Column F
Choose 2½ teaspoons of any of the following:	Choose 2½ teaspoons of any of the following:	Choose ½ teaspoon of any of the following:	Use all of the following:	Choose 80 ml of any of the following:	Choose any of the following options:
• Apricot kernel oil • Coconut oil • Grapeseed oil • Hazelnut oil • Jojoba oil • Kukui nut oil • Sunflower oil	• Almond oil • Hemp seed oil • Meadowfoam seed oil • Pumpkin seed oil • Rosehip seed oil • Walnut oil	• Lanolin • Liquid lecithin • Vegetable glycerin	• ½ teaspoon stearic acid • 2 heaping teaspoons emulsifying wax NF	• Aloe vera gel • Distilled water • Floral water (hydrosol)	• ½ teaspoon vitamin E oil • Up to 50 drops essential oil • 1 teaspoon herbal tincture (extract)

Column A	Column B	Column C	Column D	Column E	Column F
Choose 2½ teaspoons of any of the following:	Choose 2½ teaspoons of any of the following:	Choose ¾ teaspoon of any of the following:	Choose all of the following:	Choose 70 ml of any of the following:	Choose any of the following options:
• Avocado oil • Castor oil • Coconut oil • Lanolin • Macadamia nut oil • Olive oil • Sesame oil	• Cocoa butter • Mango butter • Shea butter	• Lanolin • Liquid lecithin • Vegetable glycerin	• 1 teaspoon grated beeswax • ¾ teaspoon stearic acid • 2¼ heaping teaspoons emulsifying wax NF	• Aloe vera gel • Distilled water • Floral water (hydrosol)	• ½ teaspoon vitamin E oil • Up to 50 drops essential oil • 1 teaspoon herbal tincture (extract)

Recipes:
Six of the Best Mani/Pedi Treatments

234

Vanilla & Honey Sea-Salt Hand Scrub

Makes approx. 275 g

This luxurious scrub smells of warming and delicious vanilla with the combination of powdered vanilla beans and vanilla absolute.

200 g fine sea salt
60 ml sweet almond oil
2 tablespoons honey
1 teaspoon powdered vanilla beans
10 drops vanilla absolute

1. Combine all of the ingredients into a small glass bowl and mix well.

To use: Wash your hands and, while they are still moist, massage a tablespoon of scrub into them, using a gentle pressure, until the sugar has melted. Rinse away with warm water.

235

Blueberry & Brown Sugar Hand Scrub

Makes approx. 310 g

This recipe relies on powdered, freeze-dried blueberries, along with super-sweet brown sugar, to exfoliate, soften and polish your hands.

165 g packed dark brown sugar
3 tablespoons olive oil
2 tablespoons sweet almond oil
2 tablespoons powdered, freeze-dried blueberries

1. Combine all of the ingredients in a small glass bowl and mix well.

To use: Wash your hands and, while they are still moist, massage a tablespoon of scrub into them, using a gentle pressure, until the sugar has melted. Rinse away with warm water.

236

Cooling Peppermint Foot Mist

Makes approx. 45 ml

Mist this deodorising and cooling spray over tired feet for an instant feeling of refreshment.

¼ teaspoon cornflour
1 tablespoon witch hazel extract
1 tablespoon peppermint floral water
30 drops peppermint essential oil
15 drops tea tree essential oil

1. Add all of the ingredients to a small spray bottle and shake well. Store in the fridge and use within 2 weeks.

To use: Shake well and mist over your feet.

▶ Dried crushed thyme, honey and almonds

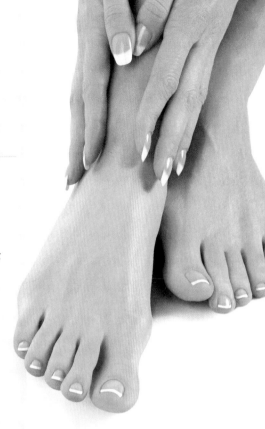

237

Herbs & Salt Reviving Foot Soak

Makes 1 application

Revive your tired toes with peppy peppermint essential oil.

275 g Epsom salts
1 teaspoon dried peppermint
1 teaspoon dried crushed thyme
1 teaspoon olive oil
10 drops peppermint essential oil
10 drops tea tree essential oil
10 drops eucalyptus essential oil
5 drops rosemary essential oil

1. Combine all of the ingredients in a small bowl and mix well.
2. Transfer the mixture to a large, cotton muslin bag and place in a large bowl or basin that will accommodate your feet.
3. Fill the container with very warm (yet comfortable) water.
4. Place your feet and ankles in the warm water and soak for 15–20 minutes.
5. Towel-dry your feet before walking.

Note: If you have a medical condition (especially diabetes, high blood pressure or any other disorder), consult a healthcare professional before performing a foot soak.

238

Tingling Tea Tree & Peppermint Foot Scrub

Makes approx. 310 g

Purify your feet with this deodorising foot scrub.

135 g fine Dead Sea salts
135 g fine Himalayan pink salt
1 tablespoon Fuller's Earth clay
50 g coconut oil (plus more if needed)
60 drops tea tree essential oil
40 drops peppermint essential oil
20 drops rosemary essential oil

1. Combine all of the ingredients in a small glass bowl and mix well. (If you need to use more coconut oil, then add this in 1 teaspoon increments until a smooth scrub forms.)

To use: Wash your feet and, while they are still moist, massage a tablespoon of scrub (using a gentle pressure) into each foot in order to deodorise and exfoliate. Leave on for up to 10 minutes. Rinse your feet well with soap and warm water. Use caution when walking, as your feet may be oily.

239

Vetiver & Patchouli Deodorising Foot Spray

Makes approx. 30 ml

Sweet-smelling feet are just a spray away with this deodorising formula!

1 tablespoon rose floral water
1 tablespoon orange blossom water
10 drops patchouli essential oil
10 drops vetiver essential oil
10 drops Australian sandalwood essential oil

1. Add all of the ingredients to a small spray bottle and shake well. Store in the fridge and use within 2 weeks.

To use: Shake well and mist over your feet for all-day-long deodorant protection.

6 Perfumes and Aromatherapy

Enter the intoxicating world of aromatherapy and perfume blending, and take your hand-crafted beauty formulas to new heights. Learn how to compose a balanced and harmonising blend with top notes, middle notes and base notes; develop that signature scent for a special someone; or create an air-diffuser aroma to fragrance your home. Discover the wonderful powers of aromatherapy oils and mix blends that will calm you, energise you, help you focus or soothe you into a restful slumber. Once you understand the basics, you will be well on your way to creating beauty products that smell every bit as good as they feel.

Perfumes and Aromatherapy

The art of blending natural perfumes is comparable to arranging flowers in an exquisitely perfumed bouquet. There is no strict science or wrong or right approach behind blending perfumes; it is an art that is improved through experimentation and practice. However, there are several 'trade secrets', that will help you on your way to creating beautiful natural perfumes. Check out the fragrance wheel on page 138 as a handy resource.

Check out the fragrance wheel on page 138 as a handy resource.

FIX IT

Many absolutes are thick and viscous, which can make them hard to work with in their undiluted state. Many companies offer pre-diluted absolutes that are easier to blend into oil-based perfumes and these will be less expensive compared with their undiluted versions.

240

Understanding top, middle and base notes

Essential oils, absolutes and CO_2 extracts are categorised as top, middle or base notes, depending on the rate of evaporation of the oil.

Top notes: Make up 5% to 20% of the perfume blend. The oils in this category are those that evaporate the most quickly and the ones you will experience first when taking a sniff of a perfume blend. The aromas of these notes are 'crisp', 'insistent' and 'sharp'.

Middle notes: Make up 50% to 80% of the perfume blend. The oils in this category help to synchronise the perfume blend and do not evaporate as rapidly as the top-note oils. These oils are considered the 'heart' of the perfume blend.

Base notes: Make up 5% to 20% of the perfume blend. The oils in this category serve as a fixative for the perfume blend and help to extend the amount of time it remains aromatic on the skin. These particular oils have an unhurried evaporation rate and are typically thick and viscous. They can be a challenge to work with at times. Your nose will generally become aware of these aromas upon the dry-down, or about 20 minutes after you apply the perfume.

241

Saving sandalwood

Sandalwood (*Santalum album*) from the Mysore region of India is almost extinct due to reckless overharvesting and should not be used if found. More sustainable choices when choosing sandalwood essential oil are Australian sandalwood, New Caledonia sandalwood or Hawaiian sandalwood.

Note: Some of the essential oils, absolutes and CO_2 extracts listed in the chart on the right may be photosensitising or unsuitable for small children, pregnant women and breast-feeding women, as well as for those with medical conditions and/or taking medications. These oils are highly concentrated and need to be handled with respect and care. Never use them undiluted on the skin. It is suggested that you research specific oils thoroughly before using them to determine the precise safety guidelines for each one.

242

Commonly used ingredients in perfume blending

This chart lists some of the essential oils, absolutes and CO_2 extracts most commonly used in perfume blending. The oils, absolutes and extracts are arranged by their note classification (top, middle or base; see left) and you will also find key words for their aroma to help you choose the right scents for your blend.

Note Classification	Aroma Notes	Essential Oil/Absolute/Extract
Top note and Middle note	Woody, spicy	**Coriander seed** (*Coriandrum sativum*)
Top note and Middle note	Woody, sweet	**Cypress, blue** (*Callitris columellaris*)
Top note	Herbaceous, lemon	**Lemon tea tree** (*Leptospermum petersonii*)
Top note and Middle note	Herbaceous, sweet	**Melissa** (*Melissa officinalis*)
Top note and Middle note	Sweet, spicy	**Basil** (*Ocimum basilicum*)
Top note and Middle note	Sweet, spicy	**Fennel, sweet** (*Foeniculum vulgare*)
Top note and Middle note	Sweet, floral, orange-like	**Mandarin, red** (*Citrus reticulate*)
Top note	Sweet, orange citrus	**Orange, sweet** (*Citrus sinensis*)
Top note	Sweet, minty	**Spearmint** (*Mentha spicata*)
Top note	Floral, sweet	**Geranium, rose** (*Pelargonium roseum*)
Top note and Middle note	Floral, fresh, sweet, herbaceous	**Lavender** (*Lavandula angustifolia*)
Top note	Floral, soft	**Palmarosa** (*Cymbopogon martinii*)
Top note	Floral, rich, sweet	**Ylang ylang absolute** (*Cananga odorata*)
Top note	Floral, rich, sweet	**Ylang ylang extra** (*Cananga odorata*)
Top note	Floral, rich, sweet	**Ylang ylang I** (*Cananga odorata*)
Top note	Floral, rich, sweet	**Ylang ylang II** (*Cananga odorata*)
Top note	Floral, rich, sweet	**Ylang ylang III** (*Cananga odorata*)
Top note	Fresh, green	**Galbanum** (*Ferula galbaniflua*)
Top note and Middle note	Fresh, lemon, sweet	**Lemongrass** (*Cymbopogon flexuosus*)
Top note	Fresh, minty, sweet	**Peppermint** (*Mentha piperita*)
Top note	Fresh, green, tart	**Verbena** (*Lippia citriodora*)
Top note	Citrus, sweet, green	**Bergamot** (*Citrus aurantium var. bergamia*)
Top note	Citrus, green	**Citronella** (*Cymbopogon nardus*)
Top note	Citrus, fresh	**Eucalyptus, lemon** (*Eucalyptus citriodora*)
Top note	Citrus, sweet	**Grapefruit, pink** (*Citrus paradisi*)
Top note	Citrus, sweet	**Grapefruit, ruby red** (*Citrus paradisi*)
Top note	Citrus, sweet	**Grapefruit, white** (*Citrus paradisi*)
Top note	Citrus, fresh	**Lemon** (*Citrus limon*)
Top note	Citrus, sweet, fresh	**Lime (distilled)** (*Citrus aurantifolia*)
Top note	Citrus, fruity, tart	**Lime (expressed)** (*Citrus aurantifolia*)
Top note	Citrus, sweet, green	**Orange, bitter** (*Citrus aurantium var. amara*)
Top note	Citrus, sweet	**Orange, blood** (*Citrus sinensis*)
Top note and Middle note	Citrus, sweet, warm	**Tagetes** (*Tagetes bipinata*)
Top note	Citrus, zesty, lemony	**Yuzu** (*Citrus junos*)

TOP NOTES

MIDDLE NOTES

Note Classification	Aroma Notes	Essential Oil/Absolute/Extract
Middle note and Base note	Woody, green	**Angelica root CO$_2$** (*Angelica archangelica*)
Middle note and Base note	Woody, sweet	**Cedarwood, Virginia** (*Juniperus virginiana*)
Middle note and Top note	Woody, sweet	**Cypress, blue** (*Callitris columellaris*)
Middle note	Woody, warm	**Ginger lily** (*Hedychium spicatum*)
Middle note	Woody, fresh	**Juniper berry CO$_2$** (*Juniperus communis*)
Middle note	Woody, sweet	**Palo santo** (*Bursera graveolens*)
Middle note	Woody, green, citrus	**Petitgrain absolute** (*Citrus aurantium*)
Middle note	Earthy, herbaceous	**Helichrysum** (*Helichrysum italicum*)
Middle note	Earthy, floral, green	**Lotus, pink absolute** (*Nelumbo nucifera*)
Middle note	Earthy, floral, green	**Lotus, white absolute** (*Nelumbo nucifera*)
Middle note	Herbaceous, green	**Calendula CO$_2$** (*Calendula officinalis*)
Middle note	Herbaceous, sweet	**Clary sage** (*Salvia sclarea*)
Middle note	Herbaceous, fresh	**Cypress leaf** (*Cupressus sempervirens*)
Middle note	Herbaceous, warm	**Davana** (*Artemisia pallens*)
Middle note	Herbaceous, green	**Geranium absolute** (*Pelargonium x asperum*)
Middle note	Herbaceous, warm, spicy	**Marjoram** (*Origanum majorana*)
Middle note	Herbaceous, fresh	**Myrtle, green** (*Myrtus communis*)
Middle note	Herbaceous, spicy	**Rosemary** (*Rosmarinus officinalis*)
Middle note	Herbaceous, warm	**Sage** (*Salvia officinalis*)
Middle note and Base note	Herbaceous, warm	**Seaweed absolute** (*Fucus vesiculosus*)
Middle note and Top note	Sweet, spicy	**Basil** (*Ocimum basilicum*)
Middle note	Sweet, beeswax and honey	**Beeswax absolute**
Middle note	Sweet, fruity, slightly earthy	**Blackcurrant bud absolute** (*Ribes nigrum*)
Middle note	Sweet, herbaceous, warm	**Chamomile, blue** (*Matricaria chamomilla*)
Middle note	Sweet, fruity	**Chamomile, Roman** (*Anthemis nobilis*)
Middle note	Sweet, spicy	**Clove bud** (*Eugenia caryophyllata*)
Middle note and Top note	Sweet, spicy	**Fennel, sweet** (*Foeniculum vulgare*)
Middle note	Sweet, balsamic	**Fir, balsam** (*Abies balsamea*)
Middle note and Top note	Sweet, lemon, fresh	**Lemongrass** (*Cymbopogon flexuosus*)
Middle note	Sweet, floral, orange-like	**Mandarin, red** (*Citrus reticulate*)

BASE NOTES

Note Classification	Aroma Notes	Essential Oil/Absolute/Extract
Base note	Woody, smoky	**Amber oil** (*Oleum succini /Anbar*)
Base note	Woody	**Amyris** (*Amyris balsamifera*)
Base note	Woody, green	**Angelica root CO$_2$** (*Angelica archangelica*)
Base note	Woody, warm	**Cedarwood, Atlas** (*Cedrus atlantica*)
Base note and Middle note	Woody, sweet	**Cedarwood, Virginia** (*Juniperus virginiana*)
Base note	Woody, sweet	**Frankincense** (*Boswellia carterii*)
Base note	Woody, warm	**Sandalwood** (*Santalum album*) See page 128
Base note	Woody, warm, sweet	**Sandalwood, Hawaiian** (*Santalum paniculatum*)
Base note	Woody, warm, sweet	**Sandalwood, Australian** (*Santalum spicatum*)
Base note and Middle note	Sweet, vanilla-like	**Balsam of Peru** (*Myroxylon pereirae*)
Base note	Sweet, soft, vanilla-like	**Benzoin resin oil** (*Styrax tonkinensis*)
Base note	Sweet, green	**Cassie absolute** (*Acacia farnesiana*)
Base note and Middle note	Sweet, wine-like aroma	**Cognac, green** (*Vitis vinifera*)
Base note	Sweet, warm, deep	**Labdanum absolute** (*Cistus ladaniferus*)
Base note	Sweet, smoky, warm	**Tabacco absolute** (*Nicotiana tabacum*)
Base note	Sweet, rich, warm	**Vanilla absolute** (*Vanilla planifolia*)
Base note	Sweet, rich, warm	**Vanilla bourbon CO$_2$** (*Vanilla planifolia*)
Base note	Earthy, woody	**Agarwood CO$_2$** (*Aquilaria agallocha*)

Note Classification	Aroma Notes	Essential Oil/Absolute/Extract
Middle note and Top note	Sweet, herbaceous	**Melissa** (*Melissa officinalis*)
Middle note	Sweet, earthy	**Petitgrain, mandarin** (*Citrus reticulate*)
Middle note	Sweet, floral, rosy	**Rose otto** (*Rosa damascena*)
Middle note and Top note	Sweet, warm, citrus	**Tagetes** (*Tagetes bipinata*)
Middle note	Sweet, vanilla-like	**Tonka bean absolute** (*Dipteryx odorata*)
Middle note	Floral, spicy	**Champaca CO$_2$** (*Michelia champaca*)
Middle note and Base note	Floral, sweet	**Jasmine absolute** (*Jasminum grandiflorum*)
Middle note and Base note	Floral, sweet	**Jasmine CO$_2$** (*Jasminum grandiflorum*)
Middle note	Floral, sweet	**Jasmine concrete** (*Jasminum grandiflorum*)
Middle note	Floral, sweet	**Jasmine sambac absolute** (*Jasminum sambac*)
Middle note and Top note	Floral, fresh, sweet, herbaceous	**Lavender** (*Lavandula angustifolia*)
Middle note	Floral, green, fresh	**Lavender absolute** (*Lavandula angustifolia*)
Middle note	Floral, deep	**Orange blossom absolute** (*Citrus aurantium*)
Middle note	Floral, intense, rosy	**Rose absolute** (*Rosa damascena*)
Middle note	Floral, soft, rosy	**Rose de mai absolute** (*Rosa centifolia*)
Middle note and Base note	Floral, fruity, sweet	**Tuberose absolute** (*Polianthes tuberosa*)
Middle note	Spicy, woody	**Caraway CO$_2$** (*Carum carvi*)
Middle note	Spicy, fresh, sweet	**Cardamom** (*Elettaria cardamomum*)
Middle note	Spicy, floral, green	**Carnation absolute** (*Dianthus caryophyllus*)
Middle note	Spicy, warm	**Cinnamon bark** (*Cinnamomum zeylanicum*)
Middle note and Top note	Spicy, woody	**Coriander seed** (*Coriandrum sativum*)
Middle note	Spicy, fresh, citrus	**Elemi** (*Canarium luzonicum*)
Middle note	Spicy, warm	**Ginger** (*Zingiber officinale*)
Middle note	Spicy, fresh	**Pepper, black** (*Piper nigrum*)
Middle note	Spicy, sweet	**Peppercorn, pink** (*Schinus molle*)
Middle note	Rich, butter aroma	**Butter CO$_2$**
Middle note	Rich, chocolate aroma	**Cocoa absolute** (*Theobroma cacao*)
Middle note	Rich, coffee aroma	**Coffee bean oil** (*Coffea arabica L.*)

Note Classification	Aroma Notes	Essential Oil/Absolute/Extract
Base note	Earthy, resinous	**Myrrh** (*Commiphora myrrha*)
Base note	Earthy, mossy, deep	**Oakmoss absolute** (*Evernia prunastri*)
Base note	Earthy, rich, deep, sweet, intense	**Patchouli** (*Pogostemon cablin*)
Base note	Earthy, rich, deep, sweet, intense	**Patchouli CO$_2$** (*Pogostemon cablin*)
Base note	Earthy, sweet, woody	**Spikenard** (*Nardostachys jatamansi*)
Base note	Earthy, smoky, dark	**Vetiver** (*Vetiveria zizanioides*)
Base note and Middle note	Floral, sweet	**Jasmine absolute** (*Jasminum grandiflorum*)
Base note and Middle note	Floral, sweet	**Jasmine CO$_2$** (*Jasminum grandiflorum*)
Base note	Floral, sweet, fruity	**Osmanthus absolute** (*Osmanthus fragrans*)
Base note and Middle note	Floral, sweet, fruity	**Tuberose absolute** (*Polianthes tuberosa*)
Base note	Floral, soft, green	**Violet leaf absolute** (*Viola odorata*)
Base note	Herbaceous, warm	**Cistus** (*Cistus ladaniferus*)
Base note	Herbaceous, sweet	**Hay absolute** (*Foin coupe*)
Base note and Middle note	Herbaceous, salty, mossy	**Seaweed absolute** (*Fucus vesiculosus*)
Base note	Musky	**Ambrette seed CO$_2$** (*Hibiscus abelmoschus*)

243

Using the tools of the trade

There are some essential tools that will come in handy when you start formulating perfume blends, as follows:

Small glass bottles: Available in sizes of 2–4 ml with droppers or caps, these small bottles are wonderful for storing undiluted blends while they mature. Choose amber glass or cobalt blue when available. **Tip:** Avoid storing citrus oils in bottles that have rubber droppers, as the citrus oils can ruin the rubber material.

Disposable pipettes: These are available in both 1 ml and 3 ml sizes. They are used to measure liquids by the drop and by volume.

Scent strips: These thin strips of paper are used when composing a perfume blend. You can place single drops of oil on the end of individual test strips to evaluate the aromas of the oils by themselves or next to other loaded test strips to gain an idea of how multiple oils will interact and harmonise with each other. They are usually sold in packs of 100.

Decorative perfume bottles: You can store diluted perfume blends in vials, roll-on bottles, decorative perfume bottles, perfume atomisers and jars or tins for solid perfume blends. You will find various sizes available, ranging from 5 ml up to 60 ml.

244

Creating and storing perfumes

As soon as you have composed a perfume blend, allow it to mature and ripen for at least 10 days before diluting with a carrier. You may be pleasantly surprised at how the blend transforms and find it aromatically amazing. You will then need to store or dilute the perfume blend, following the guidelines given below:

Storing perfume blends: For the best results, store your blend in an amber- or dark-coloured glass bottle with a tight-fitting lid. The blend should be kept away from heat and direct sunlight.

Diluting perfume blends: You may choose to dilute your perfume blend in an odourless carrier oil, such as jojoba, fractionated coconut, sunflower or vegetable oil. You can also use perfumers' alcohol, which is a high-proof ethanol containing denaturants. If you are using your blend strictly for yourself, you may choose to use 190-proof vodka that can be purchased from some off licences. You can also make a solid perfume by diluting your perfume blend in an oil carrier and thickening it with beeswax. To prevent skin sensitivities to the essential oils, absolutes and/or CO_2 extracts in the perfume blend, you should use no more than 10% of the blend in your chosen carrier. For example, if you are making a 15-ml bottle of perfume, you should only add up to 40 drops of your perfume blend to the carrier oil or alcohol base.

245

3

Make a basic solid perfume

Ingredients:
• *1 tablespoon jojoba, fractionated coconut or sunflower seed oil (this is the carrier oil)*
• *2 teaspoons grated beeswax*
• *50–55 drops of perfume blend*
Makes approx. 45 ml

1. Measure the carrier oil and beeswax into a small, heatproof glass.
2. Sit the glass in a saucepan of simmering water until the oil and wax have both melted.
3. Remove from the water, add your perfume blend and stir well with a cocktail stick.
4. Carefully and immediately pour the mixture into a small jar or metal tin (with a lid).
5. Put on the lid and allow to cool until solid.

To use: Apply to skin as needed, avoiding the eyes and sensitive skin.

246

Make a basic alcohol-based perfume

Ingredients:
• *2 tablespoons perfumers' alcohol or 190-proof vodka (ethanol)*
• *40–45 drops of perfume blend*
Makes approx. 30 ml

1. Add the alcohol to a 30-ml glass spray bottle.
2. Add the perfume blend.

To use: Shake well and mist on the skin as needed, avoiding the eyes and sensitive skin.

Warning: This blend is highly flammable and should not be used near heat or an open flame. Before sending anyone an alcohol-based perfume, check with your local post office for the precise shipping rules when sending flammable liquids.

247

Make a basic oil-based perfume

Ingredients:
• *2 tablespoons jojoba, fractionated coconut or sunflower seed oil (this is the carrier oil)*
• *40–45 drops of perfume blend*
Makes approx. 30 ml

1. Place the carrier oil in a 30-ml perfume roller ball.
2. Add the perfume blend.
3. Put on the lid and shake well.

To use: Roll the perfume onto the skin as needed, avoiding the eyes and sensitive skin.

248

Take it slow

When blending perfumes, it is important to start out with just a few drops of the oils at a time. Wait a few hours and allow the oils to balance, and then sniff the blend to check how you are progressing. Add more oils, drop by drop, as you go along until you feel the blend is just right. Make sure you write down the exact perfume-blend recipe, so that you will have helpful notes for future blending.

TRY IT

Pour your solid perfume into lip-balm containers for totally portable and easy-to-apply solid perfumes.

249

Beautiful blends for oil-based perfumes

To make any of the oil-based perfumes below, blend the essential oils in a small bottle and allow to mature for at least 10 days. Dilute with 1 tablespoon of jojoba or sunflower oil in a decorative perfume bottle. Apply to pulse points as desired.

Herbal perfume
16 drops lavender essential oil
4 drops galbanum essential oil
12 drops rosemary essential oil
7 drops clary sage essential oil
1 tablespoon jojoba or sunflower oil

Earthy perfume
3 drops patchouli essential oil
3 drops vetiver essential oil
4 drops carrot seed essential oil
1 drop oakmoss absolute
1 tablespoon jojoba or sunflower oil

Floral perfume
3 drops ylang ylang essential oil
3 drops jasmine absolute
8 drops rose absolute
8 drops sandalwood essential oil
4 drops ginger lily CO_2
4 drops lavender essential oil
1 tablespoon jojoba or sunflower oil

Fruity perfume
5 drops blackcurrant bud absolute
2 drops Roman chamomile essential oil
6 drops Litsea cubeba essential oil
4 drops tangerine essential oil
2 drops bergamot essential oil
5 drops lemongrass essential oil
4 drops vanilla absolute
4 drops benzoin resin oil
1 tablespoon jojoba or sunflower oil

250

Choosing an aromatherapy oil

When you want to create different moods or feelings, choose from any of the following aromatherapy oils. Please remember that these oils need to be diluted appropriately (see page 37). This information is for educational purposes only and is not intended to diagnose, treat or cure any disease. If you have any medical condition or are pregnant or breast-feeding, you must consult a healthcare professional before using essential oils.
To achieve balance: Clary sage essential oil, rose essential oil and/or geranium essential oil
To relax: Lavender essential oil, sweet marjoram essential oil, Roman chamomile essential oil, mandarin petitgrain essential oil and/or geranium essential oil
To ignite passion: Jasmine absolute, patchouli essential oil and/or ylang ylang essential oil
To create energy: Peppermint essential oil, cypress essential oil, pine essential oil, eucalyptus essential oil, sweet orange essential oil and/or lemon essential oil
For meditation: Frankincense essential oil, myrrh essential oil, sandalwood essential oil, vetiver essential oil and/or spikenard essential oil
For restful sleep: Lavender, Roman chamomile, hops CO_2 and/or yarrow essential oil

251

How to make a freshening body spray

Ingredients:
• 120 ml rose floral water
• 7 drops rose absolute
• 4 drops lavender essential oil
• 2 drops benzoin resin oil
Makes approx. 120 ml

1. Combine all the ingredients in a spray bottle. Mix well and mist over the body to freshen up. Avoid eyes. Store in the fridge and use within 2 weeks.

TRY IT

Make an aromatherapy body massage oil by adding 15 drops of essential oil blend to 2 tablespoons of a carrier oil such as apricot, sweet almond or jojoba.

Blends for aromatic air-diffusers

Diffusing the healing properties and gorgeous aromas of essential oils into the air is easy when you use ceramic or ultrasonic essential oil diffusers. Typically, you add pure water, along with a small amount of essential oils, to the diffuser and either ignite a small candle underneath or turn on the power. Always follow the instructions for your specific diffuser to obtain the best results and to keep safe. To make a room spray, dilute 45 drops of the diffuser blend in a spray bottle containing 60 ml of distilled water. Shake the bottle well and mist into the room's air to freshen and purify.

1. Blend your choice of essential oils in a small bottle, choosing from the different blends on the right.
2. Follow the instructions for the specific diffuser on how to diffuse the oils into the air and how many drops of the diffuser blend to use.

Meditation diffuser blend
25 drops frankincense essential oil
25 drops myrrh essential oil
50 drops bergamot essential oil

Relaxing diffuser blend
25 drops sweet orange essential oil
10 drops sweet basil essential oil
10 drops Roman chamomile essential oil
25 drops lavender essential oil

Air-purifying diffuser blend
30 drops lavender essential oil
15 drops thyme essential oil
15 drops rosemary essential oil
20 drops tea tree oil
20 drops lemon essential oil

Sunny outlook diffuser blend
25 drops lemon essential oil
10 drops neroli essential oil
50 drops lavender essential oil

Winter holiday diffuser blend
20 drops clove essential oil
15 drops cinnamon bark essential oil
15 drops ginger essential oil
10 drops tangerine essential oil

Balancing your chakras

According to the philosophy behind yoga, there are seven centres of spiritual energy in the human body. A chakra represents a focus or concentration of energy in the body. 'Chakra' is the Sanskrit word for 'wheel'. Essential oils are often used during chakra yoga and/or chakra meditation to help promote a balance between the mind, body and spirit.

 Crown chakra: Vetiver, sandalwood, rosewood, rose, neroli, lavender, Helichrysum, galbanum, frankincense.

 Third eye chakra: Vetiver, sandalwood, patchouli, sweet marjoram, frankincense, elemi, clary sage.

 Throat chakra: Spearmint, peppermint, Roman chamomile, bergamot.

 Heart chakra: Rose, bergamot, geranium, cypress, lemon, neroli, ylang ylang, sandalwood.

 Solar plexus chakra: Rosemary, lemon, frankincense, myrrh, clove, juniper, lemongrass, petitgrain, spearmint, cypress, clove, cinnamon, black pepper.

 Sacral chakra: Jasmine, ylang ylang, sandalwood, cardamom, geranium, clary sage, patchouli, neroli.

 Root chakra: Vetiver, frankincense, myrrh, patchouli, ginger, angelica root.

Recipes:
Eight of the Best Aromatherapy Blends

254

Goodbye Headache Massage Blend

Makes approx. 15 ml

Find relief from headache-inducing tension with this calming blend.

1 tablespoon sweet almond oil
4 drops Helichrysum essential oil
2 drops lavender essential oil
2 drops peppermint essential oil
1 drop Roman chamomile essential oil
1 drop spearmint essential oil

1. Mix all of the ingredients in a small bottle.

To use: Apply a small amount to your face, temples, the back of your neck and chest. Breathe deeply and massage in. Avoid the eye area.

255

Sweet Slumber Insomnia Bath Treatment Blend

Makes approx. 15 ml

The sedative effects of sweet marjoram, neroli and vetiver will help you to get some much needed shut eye with this relaxing bathtime blend.

1 tablespoon double cream or whole milk
3 drops petitgrain essential oil
2 drops sweet marjoram essential oil
2 drops neroli essential oil
2 drops ylang ylang essential oil
1 drop vetiver essential oil

1. Mix all of the ingredients together well.

To use: After filling the bath with warm water, add the entire blend and swirl to mix. Enjoy the warm water, while breathing in deeply to relax. Be careful when getting out of the bath, as the surface may be slippery from the cream/milk.

256

Stop the Stress Massage Blend for Anxious Minds

Makes approx. 15 ml

Get your focus back fast with this stress-banishing massage blend with balancing rose and cheering Australian sandalwood.

1 tablespoon sweet almond oil
1 drop jasmine absolute
1 drop bergamot essential oil
1 drop sweet orange essential oil
1 drop rose essential oil
2 drops ylang ylang essential oil
2 drops Australian sandalwood essential oil

1. Mix all of the ingredients in a small bottle.

To use: Apply a small amount to your face, temples, the back of your neck and chest. Breathe deeply and massage in. Avoid the eye area.

◀ Dried lavender buds, almonds and rosemary

► Rose and eucalyptus

257

Pick-Me-Up Blend for Sadness

Makes approx. 15 ml

Feeling blue? Turn your frown upside down and put some cheer into your heart with this happy-in-a-bottle blend.

1 tablespoon sweet almond oil
2 drops vanilla absolute
1 drop jasmine absolute
2 drops rose essential oil
2 drops ylang ylang essential oil

1. Mix all of the ingredients in a small bottle.

To use: Apply a small amount to your face, temples, the back of your neck and chest. Breathe deeply and massage in. Avoid the eye area.

258

Help Me Remember! For Focus and Good Memory

Makes approx. 15 ml

Massage this blend of stimulating essential oils into your temples before studying or taking a test and your mind will be as sharp as a tack.

1 tablespoon sweet almond oil
3 drops rosemary essential oil
3 drops lemon essential oil
1 drop juniper essential oil
1 drop ginger essential oil
2 drops clove essential oil

1. Mix all of the ingredients in a small bottle.

To use: Apply a small amount to your face, temples, the back of your neck and chest. Breathe deeply and massage in. Avoid the eye area.

259

Air Disinfectant Blend

Makes approx. 30 ml

Purify your personal space with just a few mists of this germ-busting blend.

2 tablespoons rubbing alcohol
6 drops red thyme essential oil
6 drops eucalyptus essential oil
5 drops lemon essential oil
5 drops tea tree essential oil
5 drops rosemary essential oil

1. Put the rubbing alcohol in a spray bottle and add the essential oils.

To use: Spray this air-purifying blend around your personal space whenever needed. It is flammable, so avoid spraying near heat sources.

260

Eucalyptus Chest Rub

Makes approx. 30 ml

Breathe better with this eucalyptus-based blend! Not to be used on small children.

2 tablespoons shea butter (at room temperature)
10 drops eucalyptus essential oil
4 drops fir needle essential oil
2 drops myrtle essential oil
6 drops lavender essential oil

1. Mix all of the ingredients in a small, lidded container.

To use: Rub a generous amount onto your chest and neck, and breathe in deeply.

261

Yoga Massage Oil

Makes approx. 60 ml

The grounding and balancing oils in this blend will help you focus better during your yoga sessions.

4 tablespoons sweet almond or jojoba oil
8 drops frankincense essential oil
7 drops cedarwood Atlas essential oil
2 drops cistus essential oil
2 drops myrrh essential oil
5 drops Australian sandalwood essential oil

1. Mix all of the ingredients in a bottle with lid.

To use: Massage generously over your body before a yoga session.

Resources

Books

Worwood, Valerie Ann *The Complete Book of Essential Oils and Aromatherapy*

Schnaubelt, Kurt Ph.D. *The Healing Intelligence of Essential Oils: The Science of Advanced Aromatherapy*

Tisserand, Robert and Rodney Young *Essential Oil Safety: A Guide for Health Care Professionals*

Gladstar, Rosemary *Rosemary Gladstar's Medicinal Herbs: A Beginner's Guide: 33 Healing Herbs to Know, Grow and Use*

Retailers

Aromatics International
aromaticsinternational.com

Avery
avery.com

Eden Botanicals
edenbotanicals.com

G Baldwin & Co.
baldwins.co.uk

Just A Soap
justasoap.co.uk

Lotion Crafter
lotioncrafter.com

Moo
moo.com

Nashville Wraps
nashvillewraps.com

Now Foods
nowfoods.com

Online Labels
onlinelabels.com

Simplers Botanicals
simplers.com

SKS Bottle & Packaging
sks-bottle.com

Snow Lotus Essential Oils
snowlotus.org

Soaposh
Soaposh.co.uk

The Soap Kitchen
thesoapkitchen.co.uk

The Fragrance Wheel was developed by Michael Edwards in 1983 and is a handy chart to use when formulating fragrances. It shows where the various essential oils and absolutes fall within the standard families: floral, oriental, woody and fresh. For example, using the Fragrance Wheel, you will understand that patchouli essential oil falls into the 'Woody Oriental' category and that bergamot essential oil is classified as a 'Citrus' note.

Index

Credits

For more about Shannon Buck, please visit her blog at FreshPickedBeauty.com

Many thanks to Mountain Rose Herbs and G. Baldwin & Co. for generously supplying all the ingredients used in the book and in the development of the recipes.

PO Box 50220
Eugene, OR 97405
USA
USA Toll-Free (800) 879-3337
Outside USA (541) 741-7307
www.MountainRoseHerbs.com
customerservice@mountainroseherbs.com

171/173 Walworth Road
London SE17 1RW
UK
020 7703 5550
www.baldwins.co.uk
sales@baldwins.co.uk

The author would like to thank:

Robert Tisserand, author of *Essential Oil Safety Second Edition* (co-authored with Rodney Young) for his expert help with essential oil terminology.

Mountain Rose Herbs for the information on essential oil profiles.

Thanks to my wonderful and supporting husband and children.

Quarto would like to thank the following for supplying images for inclusion in this book:
Alex Studio, Shutterstock.com, p.13br
AlexSmith, Shutterstock.com, p.36b
Alfio, Scisetti, Shutterstock.com, p.27tr
ANCH, Shutterstock.com, p.90c
Aniszewski, Paul, Shutterstock.com, p.129
Anna, Subbotina, pp.62l, 63r
Antmagn, Shutterstock.com, p.93b
Asharkyu, Shutterstock.com, p.131
AXL, Shutterstock.com, p.125t
Barbone, Marilyn, Shutterstock.com, pp.29b, 34, 40-41t
Barna, Gyorgy, Shutterstock.com, p.127
Botamochy, Shutterstock.com, p.37
Carol.anne, Shutterstock.com, p.118l
Catherine311, Shutterstock.com, p.56bl
Corbis, pp.74, 92, 114l
Cosijn, Ysbrand, Shutterstock.com, p.69r
Daffodilred, Shutterstock.com, 119br
Dutina, Igor, Shutterstock.com, p.32, 36t
Fa Chong, Shutterstock.com, p.21tr
FomaA, Shutterstock.com, p.27br
Fragrances of the World, www.fragrancesoftheworld.com, p.138
Freya-Photographer, Shutterstock.com, p.31r
Furman, Artem, Shutterstock.com, pp.103, 120t
Getty Images, p.66b, 73
Haraldmuc, Shutterstock.com, p.26t
Hitdelight, Shutterstock.com, p.94l
Ivanova, Inga, Shutterstock.com, p.121b
Jocic, Shutterstock.com, p.61b
Jung, Christian, Shutterstock.com, p.27cl
Khorzhevska, Vita, Shutterstock.com, p.53
KK-Foto, Shutterstock.com, p.68t
Konstantin, Yuganov, Shutterstock.com, p.105
Korrr, Shutterstock.com, p.91br

Kucherova, Anna, Shutterstock.com, p.23
LianeM, Shutterstock.com, p.90bl
Macniak, Kamil, Shutterstock.com, p.96t
Malyshchyts, Viktar, Shutterstock.com, p.25
Mama Mia, Shutterstock.com, p.60
Marcinski, Piotr, Shutterstock.com, pp.84l, 85r
Miltsova, Olga, Shutterstock.com, p.76l
MJTH, Shutterstock.com, 112-113b
Natalia, Zadorozhna, Shutterstock.com, p.24
Nazzu, Shutterstock.com, p.133b
O lympus, Shutterstock.com, p.134
Oksix, Shutterstock.com, p.64l
Panda3800, Shutterstock.com, p.57
Pezzotta, Mauro, Shutterstock.com, p.21br
Reika, Shutterstock.com, p.33
Sarsmis, Shutterstock.com, p.128b
Shutoff, Linda, Shutterstock.com, pp.62t, 68b
Takayuki, Shutterstock.com, p.56tr
van der Steen, Sandra, Shutterstock.com, p.38t
Vipman, Shutterstock.com, p.123t
VladGavriloff, Shutterstock.com, p.106
Vladimira, Shutterstock.com, p.77t
Volkov, Valentyn, Shutterstock.com, pp.22, 78b
Volosina, Shutterstock.com, pp.99tr, 130
Waters, Peter, Shutterstock.com, p.93c

All step-by-step and other images are the copyright of Quarto Publishing plc.

While every effort has been made to credit contributors, Quarto would like to apologise should there have been any omissions or errors – and would be pleased to make the appropriate correction for future editions of the book.